AF717943

Titles in This Series

Volume

1 **Markov random fields and their applications,** Ross Kindermann and J. Laurie Snell

2 **Proceedings of the conference on integration, topology, and geometry in linear spaces,** William H. Graves, Editor

3 **The closed graph and P-closed graph properties in general topology,** T. R. Hamlett and L. L. Herrington

4 **Problems of elastic stability and vibrations,** Vadim Komkov, Editor

5 **Rational constructions of modules for simple Lie algebras,** George B. Seligman

6 **Umbral calculus and Hopf algebras,** Robert Morris, Editor

7 **Complex contour integral representation of cardinal spline functions,** Walter Schempp

8 **Ordered fields and real algebraic geometry,** D. W. Dubois and T. Recio, Editors

9 **Papers in algebra, analysis and statistics,** R. Lidl, Editor

10 **Operator algebras and K-theory,** Ronald G. Douglas and Claude Schochet, Editors

11 **Plane ellipticity and related problems,** Robert P. Gilbert, Editor

12 **Symposium on algebraic topology in honor of José Adem,** Samuel Gitler, Editor

13 **Algebraists' homage: Papers in ring theory and related topics,** S. A. Amitsur, D. J. Saltman, and G. B. Seligman, Editors

14 **Lectures on Nielsen fixed point theory,** Boju Jiang

15 **Advanced analytic number theory. Part I: Ramification theoretic methods,** Carlos J. Moreno

16 **Complex representations of GL(2, K) for finite fields K,** Ilya Piatetski-Shapiro

17 **Nonlinear partial differential equations,** Joel A. Smoller, Editor

18 **Fixed points and nonexpansive mappings,** Robert C. Sine, Editor

19 **Proceedings of the Northwestern homotopy theory conference,** Haynes R. Miller and Stewart B. Priddy, Editors

20 **Low dimensional topology,** Samuel J. Lomonaco, Jr., Editor

21 **Topological methods in nonlinear functional analysis,** S. P. Singh, S. Thomeier, and B. Watson, Editors

22 **Factorizations of $b^n \pm 1$, $b = 2, 3, 5, 6, 7, 10, 11, 12$ up to high powers,** John Brillhart, D. H. Lehmer, J. L. Selfridge, Bryant Tuckerman, and S. S. Wagstaff, Jr.

23 **Chapter 9 of Ramanujan's second notebook—Infinite series identities, transformations, and evaluations,** Bruce C. Berndt and Padmini T. Joshi

24 **Central extensions, Galois groups, and ideal class groups of number fields,** A. Fröhlich

25 **Value distribution theory and its applications,** Chung-Chun Yang, Editor

26 **Conference in modern analysis and probability,** Richard Beals, Anatole Beck, Alexandra Bellow, and Arshag Hajian, Editors

27 **Microlocal analysis,** M. Salah Baouendi, Richard Beals, and Linda Preiss Rothschild, Editors

28 **Fluids and plasmas: geometry and dynamics,** Jerrold E. Marsden, Editor

29 **Automated theorem proving,** W. W. Bledsoe and Donald Loveland, Editors

30 **Mathematical applications of category theory,** J. W. Gray, Editor

31 **Axiomatic set theory,** James E. Baumgartner, Donald A. Martin, and Saharon Shelah, Editors

32 **Proceedings of the conference on Banach algebras and several complex variables,** F. Greenleaf and D. Gulick, Editors

33 **Contributions to group theory,** Kenneth I. Appel, John G. Ratcliffe, and Paul E. Schupp, Editors

34 **Combinatorics and algebra,** Curtis Greene, Editor

35 **Four-manifold theory,** Cameron Gordon and Robion Kirby, Editors

36 **Group actions on manifolds,** Reinhard Schultz, Editor

37 **Conference on algebraic topology in honor of Peter Hilton,** Renzo Piccinini and Denis Sjerve, Editors

Titles in This Series

Volume

38 **Topics in complex analysis,** Dorothy Browne Shaffer, Editor

39 **Errett Bishop: Reflections on him and his research,** Murray Rosenblatt, Editor

40 **Integral bases for affine Lie algebras and their universal enveloping algebras,** David Mitzman

41 **Particle systems, random media and large deviations,** Richard Durrett, Editor

42 **Classical real analysis,** Daniel Waterman, Editor

43 **Group actions on rings,** Susan Montgomery, Editor

44 **Combinatorial methods in topology and algebraic geometry,** John R. Harper and Richard Mandelbaum, Editors

45 **Finite groups—coming of age,** John McKay, Editor

46 **Structure of the standard modules for the affine Lie algebra $A_1^{(1)}$,** James Lepowsky and Mirko Primc

47 **Linear algebra and its role in systems theory,** Richard A. Brualdi, David H. Carlson, Biswa Nath Datta, Charles R. Johnson, and Robert J. Plemmons, Editors

48 **Analytic functions of one complex variable,** Chung-chun Yang and Chi-tai Chuang, Editors

49 **Complex differential geometry and nonlinear differential equations,** Yum-Tong Siu, Editor

50 **Random matrices and their applications,** Joel E. Cohen, Harry Kesten, and Charles M. Newman, Editors

51 **Nonlinear problems in geometry,** Dennis M. DeTurck, Editor

52 **Geometry of normed linear spaces,** R. G. Bartle, N. T. Peck, A. L. Peressini, and J. J. Uhl, Editors

53 **The Selberg trace formula and related topics,** Dennis A. Hejhal, Peter Sarnak, and Audrey Anne Terras, Editors

54 **Differential analysis and infinite dimensional spaces,** Kondagunta Sundaresan and Srinivasa Swaminathan, Editors

55 **Applications of algebraic K-theory to algebraic geometry and number theory,** Spencer J. Bloch, R. Keith Dennis, Eric M. Friedlander, and Michael R. Stein, Editors

56 **Multiparameter bifurcation theory,** Martin Golubitsky and John Guckenheimer, Editors

57 **Combinatorics and ordered sets,** Ivan Rival, Editor

58 **The Lefschetz centennial conference. Part I: Proceedings on algebraic geometry,** D. Sundararaman, Editor

58.II **The Lefschetz centennial conference. Part II: Proceedings on algebraic topology,** S. Gitler, Editor

58.III **The Lefschetz centennial conference. Part III: Proceedings on differential equations,** A. Verjovsky, Editor

59 **Function estimates,** J. S. Marron, Editor

60 **Nonstrictly hyperbolic conservation laws,** Barbara Lee Keyfitz and Herbert C. Kranzer, Editors

61 **Residues and traces of differential forms via Hochschild homology,** Joseph Lipman

62 **Operator algebras and mathematical physics,** Palle E. T. Jorgensen and Paul S. Muhly, Editors

63 **Integral geometry,** Robert L. Bryant, Victor Guillemin, Sigurdur Helgason, and R. O. Wells, Jr., Editors

64 **The Legacy of Sonya Kovalevskaya,** Linda Keen, Editor

65 **Logic and combinatorics,** Stephen G. Simpson, Editor

66 **Free group rings,** Narian Gupta, Editor

The Lefschetz Centennial Conference

Proceedings on Differential Equations

SOLOMON LEFSCHETZ

CONTEMPORARY MATHEMATICS

Volume 58.III

The Lefschetz Centennial Conference

Proceedings on Differential Equations

Proceedings of The Lefschetz Centennial Conference held December 10–14, 1984

A. Verjovsky, Editor

AMERICAN MATHEMATICAL SOCIETY

Providence • Rhode Island

EDITORIAL BOARD

Irwin Kra, managing editor
Thomas F. Banchoff
M. Salah Baouendi
Daniel M. Burns
David Eisenbud
W. H. Jaco
Gerald J. Janusz
Jan Mycielski
Alan D. Weinstein

The Proceedings of The Lefschetz Centennial Conference on Algebraic Geometry, Algebraic Topology and Differential Equations was held at the Centro de Investigación y de Estudios Avanzados, in Mexico City, Mexico, December 10-14, 1984.

1980 *Mathematics Subject Classification.* Primary 58-06, 34-06, 35-06, 53-06.

Organizing committee: J. Adem. S. Gitler, J. J. Kohn, E. Ramirez de Arellano, D. Sundararaman, A. Verjovsky.

Editorial Committee of the Proceedings: A. Verjovsky, J. Adem, E. Ramirez de Arellano, S. Gitler, D. Sundararaman.

Library of Congress Cataloging-in-Publication Data
(Revised for part II)

Lefschetz Centennial Conference (1984: Mexico City, Mexico). The Lefschetz Centennial Conference.

(Contemporary mathematics, ISSN 0271-4132; v. 58)
Includes bibliographies.
Contents: pt. 1. Proceedings on algebraic geometry, D. Sundararaman, editor — pt. 2. Proceedings on algebraic topology / S. Gitler — pt. 3. Proceedings on differential equations / A. Verjovsky.
1. Algebraic geometry–Congresses. 2. Algebraic topology–Congresses. 3. Differential equations–Congresses. I. Lefschetz, Solomon, 1884-1972. II. Sundararaman, D. III. Gitler, Samuel. IV. Verjovsky A. V. Series: Contemporary mathematics (American Mathematical Society); v. 58.
QA564.L43 1984 512′.33 86-14040
ISBN 0-8218-5064-4 (pt. III)
ISBN 0-8218-5065-2 (set)

Copyright ©1987 by the American Mathematical Society. All rights reserved.
The American Mathematical Society retains all rights except those granted to the United States Government.
Printed in the United States of America.
Information on copying and reprinting can be found at the back of this volume.
This volume was printed directly from author-prepared copy.
The paper used in this book is acid-free and falls within the guidelines established to ensure permanence and durability. ∞

TABLE OF CONTENTS

PREFACE

The Solomon Lefschetz Centennial Conference was held in the Centro de Investigación y de Estudios Avanzados in Mexico City from 10 to 14, December 1984. The topics of the conference were Algebraic Geometry, Algebraic Topology and Differential Equations, the three areas in which the late S. Lefschetz had made original contributions. Experts from many countries, as well as mathematicians from Mexico, participated actively and enthusiastically resulting in a very successful conference.

This volume contains the proceedings of the conference corresponding to the general area of Differential Equations.

Financial support to the conference came from CONACyT, SEP and CINVESTAV del IPN of Mexico, and the NSF of U.S.A. cosponsored and supported the conference through their Cooperative Science Programs with Latin America and the Caribbean.

The Organizing Committee of the Conference and the Editorial Committee of the Proceedings would like to thank all the participants in the conference and all the contributors to this volume.

A. Verjovsky.

CINVESTAV, Mexico City, November 1986.

Contemporary Mathematics
Volume 58, Part III, 1987

PERTURBATION RESULTS ON THE LONG RUN BEHAVIOR OF NONLINEAR DYNAMICAL SYSTEMS

Humberto Carrillo Calvet*

ABSTRACT. The validity of the averaging approximations for ordinary differential equations in standard form is discussed. New results on the uniform validity over infinite intervals of time are presented.

1. INTRODUCTION. In this paper we will be concerned with differential equations of the form

$$\dot{x} = \varepsilon\, f(t,x,\varepsilon) \tag{E}$$

where ε is a small positive parameter. The function f is going to be considered oscillatory with respect to the variable t in a very general sense that includes periodicity as a particular case. Precisely, we will assume that $f\colon \mathbb{R} \times \mathbb{R}^n \times (0,\infty) \to \mathbb{R}^n$ is continuous, has a continuous partial derivative with respect to x and is almost periodic in t, uniformly with respect to (x,ε) in compact sets. Equations like (E) were first studied by N. Krylov and N. N. Bogolyubov who called them differential equations in *standard form*.

There is a wide spectrum of dynamical systems models where such equations appear. They are very important in Mechanics and in more general systems that present nonlinear oscillation phenomena. For instance, the equation

$$\dot{x} = f(x) + g(t/\varepsilon) \tag{1}$$

with $f\colon \mathcal{D} \subset \mathbb{R}^n \to \mathbb{R}^n$ and g a periodic vector valued function, models an autonomous dynamical system being acted on by a rapidly oscillating external forcing. Clearly the change of time scale $\tau = t/\varepsilon$ reduces equation (1) to the standard form.

Also, a system of n weakly coupled harmonic oscillators

$$\ddot{u}_k + w_k\, u_k = \varepsilon\, f_k(u_1,\dots,u_k,\dot{u}_1,\dots,\dot{u}_k) \qquad k = 1,2,\dots,n \tag{2}$$

1980 Mathematics Subject Classification: 34C29, 34C15, 34D20, 34E10.
*Partially supported by Consejo Nacional de Ciencia y Tecnología, México.

© 1987 American Mathematical Society
0271-4132/87 $1.00 + $.25 per page

can be written in the standard form. In fact, using the amplitude and phase variables $y = (y_1,\dots,y_n)$ and $\theta = (\theta_1,\dots,\theta_n)$ defined by

$$\dot{u}_k + i\, w_k \dot{u}_k = y_k\, e^{i(\theta_k + w_k t)}$$

the system (2) can be transformed into

$$\dot{\theta} = \varepsilon\, F(t,\theta,y)$$

$$\dot{y} = \varepsilon\, G(t,\theta,y).$$

2. The *Averaging Method*, notably developed by Krylov, Bogolyubov and Mitropolsky [1], consists of approximating the solutions of the equation (E) by the solution of the autonomous equation

(AE) $$\dot{\bar{x}} = \varepsilon\, f_0(\bar{x})$$

where $f_0(\bar{x})$ is the mean value of $f(t,\bar{x},\varepsilon)$ with respect to t at $\varepsilon = 0$. That is to say,

$$f_0(\bar{x}) = \lim_{T\to\infty} \frac{1}{T} \int_0^T f(t,\bar{x},0)dt.$$

Equation (AE) is called the *averaged equation* and it simplifies equation (E) by at least reducing the dimension of the system by one.

To see that this is a reasonable procedure one rescales time $(\tau = \varepsilon t)$ to obtain the equations

(E') $$\frac{dx}{d\tau} = f(\tau/\varepsilon, x, \varepsilon)$$

(AE') $$\frac{d\bar{x}}{d\tau} = f_0(\bar{x}).$$

Thus, we have that for small values of ε, the equation (E') has a very rapidly oscillating forcing term and we would expect that in the first approximation the solution, $x_\varepsilon(\tau)$, of (E) will only obey the average effect of the forcing. For the case when the right hand side of equation (E) is periodic with respect to t, the validity of this reasoning is justified by the fact that, the difference between the integral of a periodic function and its mean value tends to zero as the period tends to zero.

3. The following *decomposition theorem* (Hale [2]) is a fundamental result of the averaging theory. It says that after a change of variables the equation (E) is just a perturbation of the averaged equation. In view of this result the averaging procedure can be treated, in many respects, as a classical regular perturbation method.

THEOREM 1. Given any compact set $\Omega \subset \mathbb{R}^n$ there is an $\varepsilon_0 > 0$ and a function $u(t,y,\varepsilon)$ continuous on $\mathbb{R} \times \mathbb{R}^n \times (0,\varepsilon_0]$ such that:

i) $u(t,y,\varepsilon)$ is almost periodic in t uniformly with respect to y in compact sets for each fixed ε;

ii) $u(t,y,\varepsilon)$ has a continuous derivative with respect to t and derivatives of any arbitrary specified order with respect to y;

iii) $\varepsilon u(t,y,\varepsilon)$, $\varepsilon \frac{\partial u}{\partial y}(t,y,\varepsilon)$ tend to zero as $\varepsilon \to 0$ uniformly with respect to $t \in \mathbb{R}$ and y in compact sets;

iv) The change of variables

(3) $$x = \begin{cases} y + \varepsilon u(t,y,\varepsilon) & \text{for } (t,y,\varepsilon) \in \mathbb{R} \times \Omega \times (0,\varepsilon_0] \\ y & \text{for } (t,y,\varepsilon) \in \mathbb{R} \times \Omega \times \{0\} \end{cases}$$

transforms equation (E) into

$$\dot{y} = \varepsilon f_0(y) + \varepsilon g(t,y,\varepsilon)$$

This function, $g(t,y,\varepsilon)$, is continuous, has continuous derivative with respect to y on $\mathbb{R} \times \Omega \times [0,\varepsilon_0)$ and approaches zero as $\varepsilon \to 0$ uniformly with respect to $t \in \mathbb{R}$ and x in compact sets.

For the case in which the right hand side of the equation (E) is periodic in t, it is an easy exercise to prove that the transformation (3) with

(4) $$u(t,y,\varepsilon) = \int_0^t [f(s,y,0) - f_0(y)]ds$$

produces the desired change in equation (E).

The case when $f(t,x,\varepsilon)$ is almost periodic (a.p) is not as simple because then, the function u defined by the equation (4) is not necessarily a.p. with respect to t. It happens that the integral of an a.p. function with mean value zero is not always a.p. and in fact can be unbounded [3]. However, it can be proved that there exists a function $u(t,x,\varepsilon)$ a.p. in t uniformly with respect to x in compact sets such that

$$\frac{\partial u}{\partial t}(t,y,\varepsilon) \to f(t,y,0) - f_0(y)$$

uniformly with respect to $t \in \mathbb{R}$ and y in compact sets. Using this function u in the transformation (3), the proof of Theorem 1 follows in essentially the same way as in the periodic case.

4. In the following sections we will pay attention to the problem of the validity of the averaging method. This is the problem of justifying and giving conditions under which we can obtain information about the solutions of the full equation (E) by studying instead the averaged equation (AE). This question has been approached in different ways; there is, for instance, a class of results that guarantee the existence of almost periodic solutions of the full equation

under the assumption that the averaged equation has a stable equilibrium that satisfies a non-degeneracy condition. Here we will be interested in finding conditions under which two solutions of equations (E) and (AE) with the same initial condition approach each other as ε tends to zero, uniformly in t. We will be also interested in comparing solutions that start with nearby initial conditions.

Section 5 is devoted to the uniform validity of the approximations during finite intervals of time and the sections 6 and 7 to the validity over infinite intervals of the form $[t_0,\infty)$. The results that will be presented in Section 7 are going to be particularly useful to conclude that the averaged equation reflects the qualitative behavior of the full system.

5. It follows from Theorem 1 that the equation (E') can be transformed into

$$\frac{dy}{d\tau} = f_0(y) + g(\tau,y,\varepsilon) \tag{5}$$

with the function g continuous and $g(\tau,y,\varepsilon) \to 0$ as $\varepsilon \to 0$ uniformly for $t \in \mathbb{R}$ and x in compact sets. If $\phi(\tau)$ is a solution of (AE') defined for all positive τ, then by the continuity of the solutions of equation (5) with respect to changes in the initial conditions and the parameter ε, we have that for a given η, $T > 0$ there are positive numbers ε_0 and δ such that $0 < \varepsilon < \varepsilon_0$ and $|y_0 - \phi(0)| < \delta$ implies that the solution of (5), $y_\varepsilon(\zeta)$, with initial condition $y_\varepsilon(0) = y_0$, satisfies $|y_\varepsilon(\tau) - \phi(\tau)| < \eta$ for any $\tau \in [0,T]$. It can also be proved that because of the nature of the change of variables (3), the same is true when $y_\varepsilon(\tau)$ is a solution of (AE'). From this and taking in account that between the system (E) and the system (E') the rescaling $\tau = \varepsilon t$ is involved, we obtain the following result.

THEOREM 2. *For each positive value of* ε, *let* $\phi_\varepsilon(t)$ *be a solution of the equation* (AE). *Assume that for some* $\varepsilon > 0$, ϕ_ε *is defined for all* $t \in [0,\infty)$. *Then, given a tolerance* $\eta > 0$ *and* T *as big as we please, there are positive numbers* $\varepsilon_0(\eta,T)$ *and* $\delta(\eta,T)$ *such that* $0 < \varepsilon < \varepsilon_0$ *and* $|x_\varepsilon(0) - \phi_\varepsilon(0)| < \delta$ *implies that*

$$|x_\varepsilon(t) - \phi_\varepsilon(t)| < \eta \quad \text{for all} \quad t \in [0,T/\varepsilon].$$

Here x_ε *represents a solution of equation* (E).

This is a more general version of the Bogolyubov's theorem [4] on the validity of the averaging approximations over the so called long intervals of time. The further generality here is in the fact that we are not restricted to compare solutions of the equations (E) and (AE) that start with the same initial conditions but neighboring initial conditions are allowed.

6. The Theorem 2 justifies the application of the averaging procedure during intervals of time as long as we want by taking ε small enough. Nevertheless it has the following weakness: it could be that no matter how small we take the value of ε, there is a moment after which the approximation fails. This situation would be problematic if we are interested in the long run behavior of a dynamical system. Figure 1 illustrates this phenomenon.

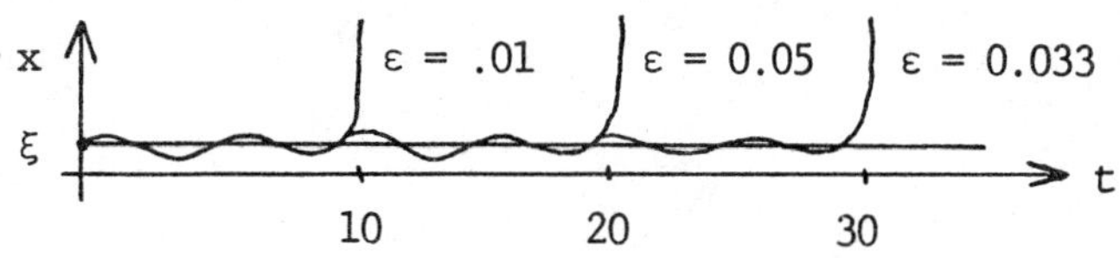

Figure 1. In the figure it is supposed that $x = \xi$ is a solution of the averaged equation and the other curves are solutions of the exact equation for different values of ε.

The system

$$\text{(E)} \qquad \begin{aligned} \dot{x}_1 &= \varepsilon\, x_2 \\ \dot{x}_2 &= -\varepsilon\, x_1 + \varepsilon \operatorname{sen} \varepsilon t \end{aligned}$$

shows us that even for solutions that are Lyapunov stable, the situation illustrated in Figure 1 can actually happen. This is a linear system with resonant forcing and has the orbit structure depicted in Figure 2(a). Averaging the right hand side we are left with the equation

$$\text{(AE)} \qquad \begin{aligned} \dot{\bar{x}}_1 &= \varepsilon\, \bar{x}_2 \\ \dot{\bar{x}}_2 &= -\varepsilon\, \bar{x}_1 \end{aligned}$$

and this has a phase portrait like the one of Figure 2(b). Clearly, the periodic solutions of (AE) can not approximate the unbounded solutions of the equation (E) uniformly for $t \in [0,\infty)$.

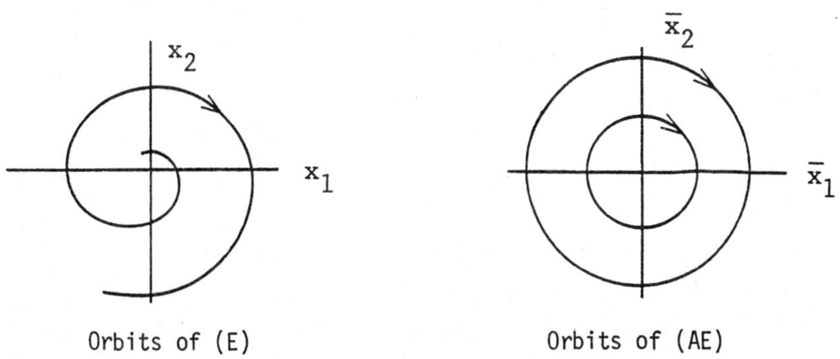

Figure 2

Once it has been established that the averaging approximation can break down during the scale of time $[t_0,\infty)$, it remains to find out under what conditions it is valid. It has been proved by Volosov [5],[6], Banfi [7] and Eckhaus [8] that for solutions with strong stability properties the approximation is justified over the unbounded interval $[t_0,\infty)$. The results that will be presented in the next section, extend and complement these previous results by treating, in a different way, more general solutions and giving us global information.

7. Let be $\phi: \mathbb{R} \times (0,\infty) \to \mathbb{R}^n$ and let us denote by x_ε a solution of equation (E). We will say that $\phi(t,\varepsilon)$ has a *stable neighborhood* (*s.n.*) as $\varepsilon \to 0$ for the equation (E), if for any $\eta > 0$ there are positive numbers $\varepsilon_0(\eta)$ and $\delta(\eta)$ such that $(t_0,\varepsilon) \in [0,\infty) \times (0,\varepsilon_0)$ and $|x_\varepsilon(t_0) - \phi(t_0,\varepsilon)| < \delta$ implies that $|x_\varepsilon(t) - \phi(t,\varepsilon)| < \eta$ for all $t \geq t_0$.

Similarly if for each $\varepsilon > 0$, γ_ε is a set in $\mathbb{R}^n$, we will say that γ_ε has an *orbitally stable neighborhood* (*o.s.n.*) as $\varepsilon \to 0$ for the equation (E) if for any $\eta > 0$ there are positive numbers $\varepsilon_0(\eta)$ and $\delta(\eta)$ such that $(t_0,\varepsilon) \in [0,\infty) \times (0,\varepsilon_0)$ and $d(x_\varepsilon(t_0),\gamma) < \delta$ implies that $d(x_\varepsilon(t),\gamma) < \eta$ for all $t \geq t_0$. Here $d(x,\gamma)$ represents the distance between the point x and the set γ.

THEOREM 3. (Ref. [9]). Let ϕ be a solution of $\dot{x} = f_0(x)$ and $x_\varepsilon(t;t_0,x_0)$ the solution of equation (E) that satisfies the initial condition $x_\varepsilon(t_0;t_0,x_0) = x_0$.

(a) If ϕ is uniformly asymptotically stable and bounded then the solution of the averaged equation, $\bar{x}_\varepsilon(t) = \phi(\varepsilon t)$, has a *s.n.* as $\varepsilon \to 0$ for equation (E). Also, if x_0 is a point in the domain of attraction of ϕ, then given $\eta > 0$, there is $\varepsilon_0(\eta) > 0$ such that for any $0 < \varepsilon < \varepsilon_0$ there exists $T(\varepsilon) > 0$ such that $|x_\varepsilon(t;t_0,x_0) - \phi(\varepsilon t)| < \eta$ for all $t \geq T$.

(b) If ϕ is orbitally uniformly asymptotically stable and bounded with orbit γ then γ has an *o.s.n.* as $\varepsilon \to 0$ for the equation (E). If x_0 is a point in the domain of attraction of γ, then given $\eta > 0$ there is $\varepsilon_0(\eta) > 0$ such that for any $0 < \varepsilon < \varepsilon_0$ there exists $T(\varepsilon) > 0$ such that $d(x_\varepsilon(t;t_0,x_0),\gamma) < \eta$ for all $t \geq T$.

Remarks. (i) The hypothesis of (a) and (b) can be satisfied for solutions that are not linearly stable. Thus, the assertions of the theorem can be also true for solutions that do not attract exponentially.

(ii) The assertion of (a) also holds when ϕ is an asymptotically stable static solution (see appendix). The same happens for (b) when ϕ is an orbitally asymptotically stable periodic solution (stable limit cycle).

iii) The statement (a) is also true for any solution stable under persistent disturbances and bounded. Furthermore, if we extend in a natural way the notion of stability when persistent disturbances for sets in $\mathbf{R}^n$ then we can obtain a similar result to that of (b) for this more general situation [10].

EXAMPLE 1. The equation

$$\text{(6)} \qquad \begin{aligned} \dot{x} &= \varepsilon y(\cos^2 t - z) \\ \dot{y} &= -\varepsilon x(1-2z + xy) \\ \dot{z} &= -\varepsilon z + \varepsilon \operatorname{sen}(t + xy) \end{aligned}$$

has not $(x,y,z) = (0,0,0)$ as a solution. However, $(0,0,0)$ is a solution of the averaged system

$$\text{(7)} \qquad \begin{aligned} \dot{x} &= \varepsilon y(\tfrac{1}{2} - z) \\ \dot{y} &= -\varepsilon x(1 - 2z + xy) \\ \dot{z} &= -\varepsilon z. \end{aligned}$$

Since $V = x^2 + \frac{1}{2}y^2 + z^2$ is a Lyapunov function for this system (there are not invariant sets in the x and y axis other than the origin) we have that $(0,0,0)$ is asymptotically stable with some domain of attraction $\mathcal{D}$. Then by Theorem 3, the origin has a stable neighborhood for equation (6), and for small ε all solutions of (6) that start on $\mathcal{D}$ will end up in a small neighborhood of the origin.

EXAMPLE 2. The planar system

$$\text{(8)} \qquad \begin{aligned} \dot{x} &= -\varepsilon y + z\varepsilon x \operatorname{sen}^2 t(1 - \sqrt{x^2+y^2})^3 \\ \dot{y} &= \varepsilon x + \varepsilon y(1 - \sqrt{x^2+y^2})^3 + \varepsilon \cos(t + x) \end{aligned}$$

has, after averaging, the polar coordinates expression

$$\text{(9)} \qquad \begin{aligned} \dot{\gamma} &= \varepsilon\gamma(1 - \gamma)^3 \\ \dot{\theta} &= \varepsilon. \end{aligned}$$

The unit circle is a stable limit cycle for this equation. Then by Theorem 3(b) it has an *o.s.n.* for equation (8). We also conclude that for small ε all solutions of (8) except the origin will be eventually within a small neighborhood of the unit circle.

APPENDIX. Let $x(t;t_0,x_0)$ denote the solution of the equation

$$\text{(10)} \qquad \dot{x} = f(t,x)$$

that satisfies the initial condition $x(t_0;t_0,x_0) = x_0$. A solution ϕ of (10) is called *uniformly stable* if it is defined for $t \geq 0$ and given any $\eta > 0$ there is $\delta(\eta)$ such that for any $t_0 \geq 0$, $|x_0 - \phi(t_0)| < \delta$ implies that $|x(t;t_0,x_0) - \phi(t)| < \eta$ for all $t \geq t_0$. ϕ is called *uniformly asymptotically stable* if it is uniformly stable and there exists $h > 0$ with the property that for every $\eta > 0$ there is $T(\eta) > 0$ such that for any $t_0 \geq 0$, $|x_0 - \phi(t_0)| < h$ implies that $|x(t;t_0,x_0) - \phi(t)| < \eta$ for $t \geq t_0 + T$. Observe that h is independent of t_0 and T is independent of both t_0 and x_0. Let us call γ the positive semiorbit of ϕ in the phase space, that is $\gamma = \{x \in \mathbb{R}^n \mid x = \phi(t)$ for some $t \geq 0\}$. We say that ϕ is *orbitally uniformly stable* if for any $\eta > 0$ there is $\delta(\eta)$ such that for any $t_0 \geq 0$, $d(x_0,\gamma) < \delta$ implies that $d(x(t;t_0,x_0),\gamma) < \eta$ for all $t \geq t_0$. Here $d(x,X)$ represents the distance between the point $x \in \mathbb{R}^n$ and the set $X \subset \mathbb{R}^n$. If there exists also $h > 0$ with the property that for every $\eta > 0$ there is $T(\eta) > 0$ such that for any $t_0 \geq 0$, $d(x_0,\gamma) < h$ implies that $d(x(t;t_0,x_0),\gamma) < \eta$ for $t \geq t_0 + T$, then we say that ϕ is *orbitally uniformly asymptotically stable*.

If f in equation (10) is periodic in t uniformly with respect to x, it is the case that for periodic solutions of the same period as f, simple stability in the sense of Lyapunov implies uniform stability, asymptotic stability implies uniform asymptotic stability. Orbital stability implies orbital uniform stability and orbital asymptotic stability implies orbital uniform asymptotic stability [11],[12].

BIBLIOGRAPHY

1. N. N. Bogolyubov and Y. A. Mitropolsky, Asymptotic Methods in the Theory of Nonlinear Oscillations, Gordon and Breach Science Publishers, New York, 1961.

2. J. K. Hale, Ordinary Differential Equations (section v.3), Wiley Interscience, New York, 1969.

3. See appendix on almost periodic functions in reference 2.

4. See chapter on the foundations of the averaging method in reference 1.

5. V. M. Volosov, "Averaging in systems of differential equations," Russ. Math. Surveys, Vol. 17, No. 6, 1962.

6. V. M. Volosov, "Averaging on an unbounded interval," Soviet Mathematics, Translation of Doklady, Vol. III, Part II, 1962.

7. C. Banfi, "Sull'aprossimazione di processi non stationari in meccanica non lineari," Boll. Un. Mat. Italiana 22, 442-450.

8. V. Eckhaus, "New approach to the asymptotic theory of nonlinear oscillations and wave propogation," J. Math. Anal. Appl. 49, 575-611.

9. H. Carrillo Calvet, El Método de el promedio y estabilidad bajo disturbios persistentes con aplicación a sincronización racional (Doctoral Thesis), Facultad de Ciencias, UNAM, 1984.

10. H. Carrillo Calvet, "Stability under persistent disturbances and averaging," in preparation.

11. J. L. Massera, "Contributions to stability theory," Ann. of Math. 64, 1956, 182-206.

12. T. Yoshizawa, "Stability of sets and perturbed systems," Funkcial. Ekvac., 5, 1963, 31-39.

Departamento de Matematicas
Facultad de Ciencias
Universidad Nacional Autonoma de México
D.F. 04510 MEXICO

Current Address:
Lefschetz Center for Dynamical Systems
Division of Applied Mathematics
Brown University
Providence, RI 02912

Contemporary Mathematics
Volume 58, Part III, 1987

Harmonic maps and minimal surface coboundaries.

K.-C. Chang and J. Eells
(Universities of Peking and Warwick)

1. INTRODUCTION.

We describe some recent results concerning minimal surfaces; and in particular, announce a theorem on the presence of unstable minimal surfaces M in Riemannian manifolds N (see §5 below). A special case, where M is a bordered planar domain and N is Euclidean n-space $\mathbb{R}^n$, is due to Morse-Tompkins [18] and Shiffman [23, 24]. If M is a disc and $N = \mathbb{R}^n$, that special case has been recast by Struwe [25]; we shall indicate how his argument can be adapted to prove our theorem.

Several different developments in the past six years have made such an extension possible:

1) Suitable existence, uniqueness, and regularity of solutions to the Dirichlet problem; and smooth dependence on initial conditions. See §§3,5 for references.

2) Various generalizations of the proof of existence of Douglas' versions of the Plateau problem. And of the theorem of Morse-Tompkins and Shiffman. See §4.

3) A version of a deformation lemma in certain spaces of maps, in the presence of a compactness condition like that of Palais-Smale, sufficient to produce both the mountain-pass lemma and the Morse inequalities. See §5.

We thank Professor Luc Lemaire for his comments on a preliminary version of this manuscript.

2. BACKGROUND.

General references [7, 6].

Let M be a compact bordered Riemann surface with conformal structure denoted by μ ; thus M has a boundary consisting of a finite number of simple closed curves $(B_i)_{1\leq i\leq k}$. Let N be a Riemannian manifold without boundary.

© 1987 American Mathematical Society
0271-4132/87 $1.00 + $.25 per page

If $C(M,N)$ denotes the space of smooth maps $\phi: M \to N$ and $\rho: C(M,N) \to C(\partial M,N)$ the restriction map (i.e., $\rho(\phi) = \phi|_{\partial M}$), then the image $E(\partial M,N)$ of ρ consists of the union of those components of $C(\partial M,N)$ which contains a map extendable to M. Thus

$$\rho: C(M,N) \to E(\partial M,N)$$

is a locally trivial fibration carrying components of $C(M,N)$ onto components of $E(\partial M,N)$. The fibre $C_\psi(M,N)$ over a map $\psi \in E(\partial M,N)$ is the space of extensions of ψ to maps $M \to N$.

The *tension field* $\tau(\phi)$ of a map $\phi: M \to N$ is the section of $\phi^{-1}TN$ given by $\operatorname{div}(d\phi)$; in local charts (isothermal on M, arbitrary on N) that has the representation

$$\phi^\gamma_{z\bar{z}} + \Gamma^\gamma_{\alpha\beta}\, \phi^\alpha_z\, \phi^\beta_{\bar{z}} = 0 \qquad (1 \le \gamma \le n = \dim N)$$

where subscripts on ϕ denote complex derivatives and $(\Gamma^\gamma_{\alpha\beta})$ describes the Levi-Civita connexion on N .

The tension field defines a map

$$\tau: C(M,N) \to C(M,TN) \ ;$$

and $\tau^{-1}(0) = H(M,N)$, the space of *harmonic maps* $\phi: M \to N$; thus

$$\tau^{-1}(0) \cap C_\psi(M,N) = \{\phi \in H(M,N) : \phi|_{\partial M} = \psi\} \ .$$

With induced topology, $H(M,N)$ is locally compact and locally finite dimensional.

τ is the Euler-Lagrange operator of the *energy functional*

$$E(\phi) = \frac{1}{2} \int |d\phi|^2 \ .$$

The harmonic map ϕ is said to be *nondegenerate* if the Hessian $\nabla^2 E(\phi)$ is a nondegenerate quadratic form.

The *conformality operator* $\kappa: C(M,N) \to C(O^2 T^* M)$ is given by $\kappa(\phi) = (\phi^* h)^{2,0}$, where $\phi^* h$ is that symmetric 2-covariant tensor field on M obtained by pulling back via ϕ the Riėmann metric h of N, and the superscript $(2,0)$ indicates that we take the dz^2-component of its complexification. In an isothermal chart,

$$(\phi^* h)^{2,0} = \frac{1}{4}(|\phi_x|^2 - |\phi_y|^2 - 2i\langle\phi_x,\phi_y\rangle)dz^2.$$

Consequently, a map $\phi: M \to N$ is (weakly) conformal iff $\kappa(\phi) = 0$.
And *if* $\phi: M \to N$ *is harmonic, then* $(\phi^* h)^{2,0}$ *is a holomorphic quadratic differential*

on M . Also, $\tau^{-1}(0) \cap \kappa^{-1}(0)$ *is the space of minimal branched immersions of* M *in* N.

If (ϕ_0, μ_0) is a critical point of E for all conformal structures μ on M and all maps ϕ in a given relative homotopy class, then ϕ_0 is a minimal branched immersion.

3. DIRICHLET'S PROBLEM.

Assume henceforth that N is complete and satisfies the following *uniformity condition:* There are constants $c > 0$ such that any point of N is in the domain of a coordinate chart $\theta: V \to \mathbf{R}^n$ whose image is the unit disc, and

$$c|d\theta(y)Y)|^2_{\mathbf{R}^n} \leq |Y|^2_y \leq C|d\theta(y)Y|^2_{\mathbf{R}^n} \tag{3.1}$$

for any $y \in V$ and $Y \in T_y(N)$. That condition is satisfied in N is compact.

With it we have Morrey's solution [16, 17] to the Dirichlet problem:

(3.2) *For any* $\psi \in E(\partial M, N)$ *there is a harmonic map* $\phi \in C_\psi(M,N)$ *which minimizes the enery* E *on* $C_\psi(M,N)$.

Morrey states his result for planar domains M , but his proof carries over to the present situation.

We cannot always prescribe the component of $C_\psi(M,N)$, as shown by a simple example of Lemaire [14] illustrating the necessity of some hypothesis in his theorem [14, 15]:

For any $\psi \in E(\partial M,N)$ *there is an* E*-minimizing harmonic map in every component of* $C_\psi(M,N)$, *provided* $\pi_2(N) = 0$.

(3.3) Concerning uniqueness, there is a maximum principle of Jäger-Kaul [10], giving the following theorem:

Suppose $\psi(\partial M)$ *is contained in an open disc* $D(b,r) \subset N$ *with centre* b *and radius* r *and contains no geodesic biangle. If* $r < \pi/2\sqrt{B}$ *with* $B > 0$ *and the sectional curvature of* N *everywhere* $\leq B$, *then the solution in* (1.5) *and* (1.6) *is unique.*

The formulation involving no geometric biangles was made in [11].

For example, if N is a complete manifold with nonpositive sectional curvature, then in every component of $C_\psi(M,N)$ there is a unique harmonic map, which is an E-minimum in that component [9,20].

4. PLATEAU'S PROBLEM.

Now suppose that M is a compact oriented bordered surface (no conformal structure is supposed given) of genus M = p and with k boundary curves

$(B_i)_{1\leq i\leq k}$. Say that M has *type* (p,) .

On its double $\tilde{M}$ the possible *essential* (i.e., not homotopically trivial) Jordan curves are the following [15]:

1) Those contained and essential in Int M or in $\tilde{M}$-M ;
2) those intersecting two different B_i ;
3) those homotopic to a multiple or some B_i ;
4) those containing an arc with endpoints on B_i , not contractible through such arcs.

Let $C = (C_i)_{1\leq i\leq k}$ be a set of disjoint oriented Jordan curves in N ; and H a relative homotopy class of continuous L_1^2-maps ϕ: M → N carrying each B_i monotonically onto C_i (i.e., with $\phi^{-1}(y)$ connected for each $y \in C_i$) , preserving orientation ($1 \leq i \leq k$). Call such maps *admissible*.

Relative to these we define (as in |13| a *primary reduction* M' of M to be a surface (possibly disconnected) obtained from either of the following operations:

a) Take an essential Jordan curve β of Int M whose H-images are contractible in N . Cut M along β , and collapse to a point each of the resulting cut curves; or
b) take an arc η on M joining two points of some B_i which is not homotopic to an arc of B_i through arcs with endpoints on B_i , whose H-images are homotopic to arcs in C_i . Cut M along η ; the two parts of B_i , and the two copies of η form two new boundary curves. Collapse one of them to a point.

We require an *irreducibility condition* ((4.1) below), restricing the types of surfaces M which will map to coboundaries of C in N. Such a condition was first formulated and thoroughly studied by Douglas [3, 4] in case N = $\mathbb{R}^n$; and by Morrey [16, 17] in general. The present form is due essentially to Jost [13].

Define $E(C,p,H) = \inf(\liminf E(\phi_s,\mu_s))$ for all sequences of confound structures μ_s on M and continuous L_1^2-maps $\phi_s:(M,\mu_s) \to (N,h)$ inducing the same homomorphism on the fundamental group as H , and mapping $(B_i)_{1\leq i\leq k}$ admissibly onto curves $(C_{si})_{1\leq i\leq k}$ in which converge to $(C_i)_{1\leq i\leq k}$ in the Fréchet sense:

$$\inf_\alpha \max(d(\phi_s(x),\alpha(x)) : x \in B_i) \to 0 \quad \text{as} \quad s \to \infty \ ,$$

for monotone orientation preserving maps $\alpha:B_i \to C_i$ ($1 \leq i \leq k$) .

Finally, set $E^*(C,p,H) = \infty$ if p = 0 and k = 1 ; or if there are no primary reductions of M ; or

$$E^*(C,p,H) = \inf E(C,p',\alpha)$$

over all surfaces M' which are homeomorphic to primary reductions of M.

The following result is due to Morrey [16] for M a planar domain, with modifications by Jost [13] to the present form:

Let M *be a compact oriented bordered surface of type* (p,k) *; and* N *be a complete Riemannian manifold satisfying the uniformity condition. Let* $C = (C_i)_{1<i<k}$ *be a family of disjoint Jordan curves in* N *; and* H *a relative homotopy class of admissible maps* $M \to N$.

(4.1) *If* $E(C,p,H) < E^*(C,p,H)$ *, then there is a conformal structure* μ *on* M *and an admissible conformal harmonic map* $\phi:(M,\mu) \to (N,h)$ *inducing the same homomorphism of fundamental groups as* H. *Furthermore,* ϕ *can be chosen to minimize area. If* $\pi_2(N) = 0$ *, then* $\phi \in H$.

REMARK. Another sort of irreducibility is the incompressibility hypothesis that the homomorphisms

$$\pi_1(H) \ : \pi_1(M) \to \pi_1(N)$$

be injective. In that form the theorem was obtained by Lemaire [15], using ideas of Sacks-Uhlenbeck [19] and Schoen-Yau [21].

5. EXISTENCE OF UNSTABLE MINIMAL SURFACES.

We are now in a position to state the main result in [2]:

Let M be a compact oriented surface of type (p,) , and (N,h) a complete Riemannian manifold satisfying the uniformity condition (3.1). Let $C = (C_i)_{1<i<k}$ be a set of disjoint oriented Jordan curves in N satisfying the irreducibility condition (4.1). Finally, we shall require the *coeicivity condition:*

(5.1) Assume that the Dirichlet problem has a nondegenerate solution in any relative homotopy class; and if $\phi,\psi : M \to N$ are relatively homotopic harmonic maps, then (relative to an isometric embedding of N into a Euclidean space V, for convenience) we require

$$||\phi-\psi||_{C^0(M,V)} \leq \text{constant}\ ||(\phi-\psi)|_{\partial M}||_{C^0(\partial M,V)} .$$

That condition is satisfied if (N,h) has nonpositive sectional curvature, by the maximum principle of [10].

Theorem . *If* $\phi_0:(M,\mu_0) \to (N,h)$ *and* $\phi_1:(M,\mu_1) \to (N,h)$ *are homotopic admissible conformal isolated* E-*minima, then there is a conformal structure* μ *on* M *and an admissible conformal harmonic map* $\phi:(M,\mu) \to (N,h)$ *homotopic to both, which is not an* E-*minimum.*

The proof will be given in [2]. Here are the main ideas, first in case M is the unit disc D (adapted from [25]).

Step 1. As in §2, the fibration ρ has a section R over its image, given by the unique solution of Dirichlet's problem (3.2). Furthermore, the nondegeneracy of $\nabla^2 E$ in the coercivity condition insures (see [8, §4]) that R is continuous, and smooth with respect to manifold structure on suitable spaces of maps:

$$\begin{array}{c} C(D,N) \\ R \downarrow \rho \\ \text{Image } \rho \subset C(\partial D,N) \end{array} .$$

Step 2. Fix an extendible diffeomorphism $\alpha:\partial D \to C_1$. Consider the Banach space $C^0 \cap L^2_{1/2}([0,2\pi], \mathbf{R})$ with norm

$$\| \quad \| \qquad \| \quad \|_{C^0} + \| \quad \|_{L^2_{1/2}} ;$$

and let

$M = \{\psi \in C^0 \cap L^2_{1/2}([0,2\pi],\mathbf{R}) \;:\; \psi \text{ is monotone, and } \psi(\frac{2k\pi}{3}) = \frac{2k\pi}{3} ,\ k=0,1,2,3,\}$.
Then M is complete and convex.

Define the function $J:M \to \mathbf{R}$ by

$$J(\psi) = \frac{1}{2} \int_D |dR(\alpha\circ\psi)|^2 .$$

If ψ is a critical point of J with respect to M , then $R(\alpha\circ\psi)$ is conformal and harmonic. Furthermore, J satisfies (in a suitable sense) the condition of Palais-Smale with respect to M. That requires the Courant-Lebesgue lemma; and for general N the maximum principle of [10], as in (3.3).

A standard conclusion is the mountain-pass property (as well as Morse theory for functions with isolated critical points [1]), which here insures that if $\psi_0,\psi_1 \in M$ are two isolated J-minima, then there is a third critical point which is not a J-minimu.

Step 3. For arbitrary M let $R(p,k)$ be the Riemann moduli space of surfaces of type (p,k) . Then R is defined on $R(p,k) \times \text{Image } \rho$. We proceed as in Steps 1 and 2. To verify the Palais-Smale condition we need now Mumford's compactness theorem, which in the present context (as in [21,§3]) says that if (μ_k, ϕ_k) is a sequence on which E is bounded, then (μ_k) subconverges to some $\mu \in R(p,k)$.

BIBLIOGRAPHY

[1] K.-C. Chang, A variant mountain pass lemma. Scientia Sinica 26; no. 12(1983), 1241-1255. Also the lecture notes, Infinite dimensional Morse theory and its applications. 22nd. Sem. Math. Sup. Montréal (1983).

[2] K.-C. Chang and J. Eells, Unstable minimal surface coboundaries. (In preparation.)

[3] J. Douglas, Some new results in the problem of Plateau. J. Math. Phys. 15(1936), 55-64.

[4] ________, Minimal surfaces of higher topological structure. Ann. Math. 40 (1939), 205-298.

[5] R. Courant, The existence of minimal surfaces of given topological type. Acta Math. 72(1940), 51-98.

[6] ________, Dirichlet's principle, conformal mappings, and minimal surfaces. Interscience (1950).

[7] J. Eells and L. Lemaire, A report on harmonic maps. Bull. London Math. Soc. 10(1978), 1-68.

[8] ________, Deformations of metrics and associated harmonic maps. Patodi Mem. Vol. Geometry and Analysis. Tata Inst. (1981),33-45.

[9] R. Hamilton, Harmonic maps of manifolds with boundary. Springer Notes 471 (1975).

[10] W. Jäger and H. Kaul, Uniqueness and stability of harmonic maps and their Jacobi fields. manu. math. 28(1979), 269-291.

[11] J. Jost, Eine geometrische Bemerkung zur Sätzen uber harmonische Abbildungen. manu. math. 32(1980), 51-57.

[12] ________, Harmonic mappings between Riemannian manifolds. Canberra (1984).

[13] ________, Conformal mappings and the Plateau-Douglas problem.

[14] L. Lemaire, Applications harmoniques de surfaces riemanniennes. J. Diff. Geo. 13(1978), 51-78.

[15] ________, Boundary value problems for harmonic and minimal maps of surfaces into manifolds. Ann. Scuola Nor. Sup. Pisa 9 (1982), 91-103.

[16] C.B. Morrey, The problem of Plateau on a Riemannian manifold. Ann. Math. 49 (1948), 807-851.

[17] ________, Multiple integrals in the calculus of variations. Grundlehren Band 130 (1960). Springer.

[18] M. Morse, and C.B. Tompkins, The existence of minimal surfaces of general critical types. Ann. Math. 40 (1939), 443-472.

[19] J. Sacks and K. Uhlenbeck, Minimal immersion of closed Riemann surfaces. Trans. Amer. Math. Soc. 271 (1982), 639-652.

[20] R. Schoen, Existence and regularity for some geometric variational problems. Thesis, Stanford Univ. (1977).

[21] R. Schoen and S.-T. Yau, Existence of incrompressible minimal surfaces and the topology of three dimensional manifolds with non-negative scalar curvature. Ann.Math. 110 (1979), 127-142.

[22] M. Shiffman, The Plateau problem for minimal surfaces of arbitrary topological structure. Amer.J. Math. 61(1939),853-882.

[23] ________, The Plateau problem for non-relative minima. Ann.Math. 40(1939), 834-854.

[24] ________, Unstable minimal surfaces with several boundaries. Ann. Math. 43 (1942), 197-222.

[25] M. Struwe, On a critical point theory for minimal surfaces spanning a wire in $\mathbb{R}^n$. J. reine ang. Math. 349 (1984), 1-23.

Contemporary Mathematics
Volume 58, Part III, 1987

LEFSCHETZ FORMULAS FOR FLOWS

David Fried*

A major theme in dynamical systems research is the study of periodic solutions to differential equations or, in geometric language, closed orbits of flows. In Poincaré's study of the restricted 3-body problem he introduced his famous method of surface transformations where a surface of section was used to convert the closed orbits to periodic points of a twist map of an annulus. As developed by Birkhoff this became a useful way of finding closed orbits for mechanical systems with 2 degrees of freedom [B]. Even for systems of higher dimensions and for dissipative systems, Lefschetz's fixed point formula [L] provided a tool in the topological study of periodic points. Here a sum of contributions from the fixed points of a map f equals a homologically defined invariant $L(f)$, the Lefschetz number.

For many years it was hoped that for any smooth flow ϕ without rest points on a closed manifold M that the closed orbits could be correlated with the topology of M in some analogous way. One imagined taking a large system of local sections, defining return maps between parts of these sections and then, by some sort of homological bookkeeping, comparing the periodic points of these return maps to a topological invariant of the line field on M defined by ϕ. This was demolished by Wilson's plug construction in which any nonsingular flow ϕ was deformed through nonsingular flows to a ψ with no closed orbits [W]. Wilson worked with dim $M \geq 4$ and Schweitzer [Sch] extended this to dim $M = 3$. The only closed manifolds where nonsingular flows exist and where all such flows have closed orbits are the circle S^1 and the Klein bottle. In particular, there is no analogue of the Lefschetz fixed point *theorem*

$$L(f) \neq 0 \Rightarrow f \text{ has a fixed point}$$

* Partially supported by the Sloan Foundation and the National Science Foundation.

© 1987 American Mathematical Society
0271-4132/87 $1.00 + $.25 per page

for a nonsingular flow ϕ.

We will show, nonetheless, that for many significant types of ϕ there is a quantitative relation between closed orbits and the underlying topology that deserves to be called a Lefschetz *formula* for ϕ. In this one counts the closed orbits using the fixed point indices of their local Poincaré sections or, more precisely, by their Fuller indices (§4). One also takes into account the position of a closed orbit γ in $\pi_1 M$ by equipping M with a flat vector bundle E, and measuring the holonomy $\rho_E(\gamma)$ of E around the loop γ. These invariants are used to construct a zeta function $Z_{\phi,E}(s)$ of a complex variable s that we call the torsion function (§5). For a generic flow ϕ, all γ's are hyperbolic and

$$Z_{\phi,E}(s) = \prod_{\gamma} (\det I - \Delta(\gamma)\rho_E(\gamma)e^{-sl(\gamma)})^{(-1)^{u(\gamma)}}$$

where γ runs over prime closed orbits, $u(\gamma)$ is the dimension of the unstable bundle over γ and $\Delta(\gamma) = +1$ or -1 as this bundle is orientable or not. Under appropriate circumstances we will identify the special value $\Psi_{\phi,E} = |Z_{\phi,E}(0)|$ with a topological invariant of E. The quantity $-\log \Psi_{\phi,E}$ is a sort of regularized sum of the contributions of closed orbits γ, where each γ contributes the product of its Fuller index and the trace of its holonomy.

The relevant topological invariant is the R-torsion $\tau(E)$. We recall some of its history. Fifty years ago Kurt Reidemeister classified the lens spaces S^3/Γ, Γ a finite cyclic group of fixed point free orthogonal transformations, up to PL equivalence [R]. He used a certain new invariant that was quickly extended by Wolfgang Franz [Fra], who used it to classify the generalized lens spaces S^{2n+1}/Γ. This invariant was a ratio of determinants concocted from a Γ-equivariant chain complex of S^{2n+1} and a nontrivial character $\rho: \Gamma \to U(1)$ of Γ. Such a ρ determines a flat bundle E over S^{2n+1}/Γ such that E has holonomy ρ. This invariant is now called the Reidemeister torsion, or R-torsion, of E and denoted $\tau(E)$. The results of Reidemeister and Franz were extended by De Rham to spaces of constant curvature $+1$ [DR].

In all this work one used the fact that τ could distinguish homotopically equivalent spaces. Namely, given a homotopy equivalence $h: X \to Y$ the resulting isomorphism of $\pi_1 X$ and $\pi_1 Y$ sets up a correspondence of flat bundles on X and flat bundles on Y. If h is a PL equivalence, or homotopic to one, then corresponding bundles will have the same torsion but otherwise they need not [C,M3]. It is this *lack* of homotopy invariance that was valuable.

In our present applications, on the other hand, the many *positive* properties of R-torsion will be fully exploited. In §1 we will discuss these properties abstractly, postponing the very definition of R-torsion and the verification of these properties to §2. In §3 we will make some simple connections with closed orbits. In §4, §5 the torsion function is introduced. When $\Psi_{\phi,E} = \tau(E)$ we say the pair (ϕ, E) is Lefschetz, meaning that it satisfies the flow analogue of the Lefschetz fixed point formula. The results of §3 can be stated as showing that when the chain recurrent set $R(\phi)$ is circular (i.e. admits cross-sections) then for certain bundles E, (ϕ, E) is Lefschetz. The remaining sections treat geodesic flow on constant curvature spaces and geometric 3-manifolds and prove Lefschetz formulas for them. Of course these results are quite close to the original concerns of Poincaré and Birkhoff with mechanical systems. They indicate that Lefschetz's intersection theory carries over to count closed geodesics on certain spaces.

We would like to acknowledge two important influences on our work. One is the paper "Infinite Cyclic Coverings" of Milnor [M1]. Here the Lefschetz zeta function of the deck transformation group of an infinite cyclic covering was related to the R-torsion of the covering. This was a purely formal result (Milnor called it "remarkably useless") since the Lefschetz numbers he considered were for deck transformations that have no fixed points. As will be seen below, in the special case of the covering associated to a cross-section to a flow ϕ this formal result has meaning as a Lefschetz formula for the closed orbits of ϕ.

The paper "Knots, Links and Symbolic Dynamics" [Fr] by Franks showed that the Alexander polynomial of a link in S^3 can be calculated from the closed orbits of certain Smale flows on the exterior of the link. By the result of [M4] identifying the Alexander polynomial with R-torsion, one sees that this is a Lefschetz formula for these flows in the sense used above: R-torsion is calculated from closed orbits. This paper directly influenced our work in [F1], where Franks' results were generalized as described in §3.

One point should be made about two ways of viewing R-torsion for line bundles E. If one wishes to consider all line bundles at once, i.e. all representations $\pi_1 \to U(1)$, one may work with the abelianization H_1 of π_1 and set up a sort of universal torsion invariant that we denote ALEX [F1]. This is a formal ratio of elements of the integral group ring $\mathbf{Z}H_1$, that directly generalizes the Alexander polynomial of a link. To obtain $\tau(E)$ from ALEX one changes rings from $\mathbf{Z}H_1$ to $\mathbf{C}$ via the ring homomorphism induced by the multiplicative homomorphism $H_1 \to \mathbf{C}^*$ induced by the holonomy representation ρ_E. In the papers of

Milnor and Franks cited above, as in [F1], one used ALEX, not $\tau(E)$. The relation between ALEX and $\tau(E)$ is that of a rational function as a formal ratio of polynomial expressions and the corresponding rational function of several complex variables, i.e. the algebraic and analytic aspects of a rational function. Only $\tau(E)$ will be used in this paper, so that we can use π_1 more fully and extract more information when π_1 is nonabelian.

We heartily thank the organizers of the Lefschetz Conference for their energy and patience.

§1. R-torsion counts circles

We will begin to develop the theme that periodic orbits of a suitable flow can be counted with R-torsion. Our methods in the first few sections will be drawn from algebraic topology, especially the Lefschetz fixed point formula.

For motivation we will recall the simpler situation of counting the stationary points of a flow with Euler characteristic. Let M be a closed smooth manifold, X a smooth vector field on M, ϕ_t the flow generated by X. Then X may be viewed as a section of the tangent bundle TM with image σ_X. The intersection number μ of this section with the image σ_0 of the zero vector field (i.e. the zero section of TM) can be computed near the set $Z_X \subset M$ where X is zero. Of course Z_X is just the stationary points of the flow ϕ_t. Now μ is independent of X, since any two σ_X are isotopic in TM. By choosing X to be the gradient vector field of a Morse function on M, one finds that μ equals the Euler characteristic $\chi(M)$ [M2].

When p is an isolated point of Z_X its contribution to μ is called the Poincaré-Hopf index $\mathrm{ind}_{PH}(X,p)$. It is computed in local coordinates as follows. Say $U \subset M$ is an open neighborhood of p that contains no other zero of X and $\psi: U \to \mathbf{R}^d$, $d = \dim M$, is a chart with $\psi(p) = 0$. If ϵ is small then the ball B_ϵ of radius ϵ centered at 0 lies in $\psi(U)$. The vector field $\psi_* X$ on $\psi(U)$ can be identified with a map $Y: \psi(U) \to \mathbf{R}^d$ in the usual way. One has a map $f: S^{d-1} \to S^{d-1}$ defined by $f(v) = -Y(\epsilon v)/\|Y(\epsilon v)\|, v \in S^{d-1}$. The degree of f is $\mathrm{ind}_{PH}(X,p)$.

With this local calculation, one obtains the theorem of Poincaré-Hopf:

For X a smooth vector field on a smooth closed manifold M with finitely many zeroes p

$$\chi(M) = \sum_p \mathrm{ind}_{PH}(X,p)$$

where $\mathrm{ind}_{PH}(X,p)$ is the local degree of $-X$ near p.

One has the same formula when M has boundary and X points inward on ∂M [M2].

It is not a coincidence that χ occurs in the above formula. Of all the homotopy invariants of a finite complex, χ is the only one that satisfies the "cardinality law" for subcomplexes

$$\chi(A \cup B) = \chi(A) + \chi(B) - \chi(A \cap B)$$

and the normalization $\chi(pt.) = 1$ (c.f. [Wa]). Thus $\chi(A)$ is the "homotopy cardinality" of the finite complex A. So χ ought to count points and the Poincaré-Hopf theorem gives a case where it does. Another case is given by the Gauss-Bonnet theorem where $2\pi\chi$ counts the total curvature of a closed Riemannian surface: one may regard this integral as a weighted sum over all the points in the surface.

There is another topological invariant for which a weakened version of the above cardinality law holds, namely R-torsion. It then might be expected that this invariant counts something geometric: we will show that in many cases it counts the periodic orbits of a flow.

R-torsion is defined in the following setting. Suppose K is a finite complex and E is a flat complex vector bundle of finite dimension with base K. Let $H^i(K;E)$ be the i-dimensional cohomology of K with coefficients in E, i.e. the twisted cohomology of E. Suppose one has a positive density on each fiber of E that is flat, i.e. locally constant in K (recall that a positive density on a complex vector space V of dimension n is a nonzero map $D: \Lambda^n V \to [0,\infty)$ satisfying $D(\lambda\omega) = |\lambda| D(\omega)$, where $\lambda \in \mathbf{C}, \omega \in \Lambda^n V$). Suppose also that $H^i(K;E)$ has a given positive density D_i for each i. Then the R-torsion $\tau = \tau(E, D_i)$ is a certain positive real number depending on E and the D_i. We postpone its definition to section 2. It satisfies a simple dependence on the D_i, namely

$$\tau(E, a_i D_i) = \tau(E, D_i) \prod_j a_j^{(-1)^{j+1}}, \quad a_j \in (0,\infty)$$

Also, τ is multiplicative under direct sums of flat bundles.

A case of particular interest is when E is an acyclic bundle, meaning that the twisted cohomology of E is zero. Then one of course can take D_i to be the absolute value map on $\Lambda^o(0) = \mathbf{C}$ and the resulting R-torsion is denoted by $\tau(E)$. In this case it does not depend on the choice of flat density on E.

We recall that a flat vector bundle over K is essentially the same as a representation of $\pi_1 K$, when K is connected. If $p \in K$ is a basepoint then one can move the fiber at p in a locally constant way around a loop in K to define an action of $\pi = \pi_1(K,p)$ on the fiber E_p of E above p. This is the holonomy representation $\rho_E: \pi \to Gl(E_p)$. Conversely, suppose one is given a representation $\rho: \pi \to Gl(V)$, V a finite dimensional complex vector space. One defines a flat bundle $E = E_\rho = \tilde{K} \times V/\pi$, where $\tilde{K}$ is the universal cover of K

and π acts on $\tilde{K}$ by deck transformations and on V by ρ. The holonomy of E_ρ is ρ, so the two constructions give an equivalence of flat bundles and representations of π.

If K is not connected (or not known to be connected, e.g. if K is defined as an intersection) it is better to work with flat bundles. One can still define holonomy using the direct sum of π_1 of the components of K as $\pi_1 K$. Using the right notion of universal cover, one again can convert a representation ρ of $\pi_1 K$ into a flat bundle E_ρ with holonomy ρ, and recover the equivalence that comes so easily in the connected case. So one may speak of the twisted cohomology of a representation ρ of $\pi_1 K$ (meaning the cohomology of E_ρ) or of an acyclic ρ (meaning that E_ρ is acyclic). In terms of ρ, our assumption that E_ρ has a flat density is just $|\det \rho| = 1$.

The R-torsion of an acyclic bundle E has many nice properties. Suppose A is a subcomplex of K and write $E|A$ for the restriction of E to A. Then we have a multiplicative cardinality law

$$(I) \qquad \tau(E|A \cup B)\tau(E|A \cap B) = \tau(E|A)\tau(E|B)$$

that is interpreted as follows. If three of the bundles $E|A \cup B$, $E|A \cap B$, $E|A$ and $E|B$ are acyclic then so is the fourth and the equation (I) holds.

Another property is the simple homotopy invariance of τ. Suppose K' is a subcomplex of K obtained by an elementary collapse of an n-cell σ in K. This means that $K = K' \cup \sigma \cup \sigma'$ where σ' is an $(n-1)$ cell of K so set up that $\partial\sigma' = \sigma' \cap K'$ and $\sigma' \subset \partial\sigma$, i.e. σ' is a free face of σ. So one can push σ' through σ into K' giving a homotopy equivalence. Then there is an obvious 1-1 correspondence of flat bundles on K and K' that doesn't change the twisted cohomology, i.e. $H^*(K; E) = H^*(K'; E|K')$. One has as well

$$(II) \qquad \tau(E, D_i) = \tau(E|K', D_i)$$

By iterating a sequence of elementary collapses and their inverses, one obtains a homotopy equivalence of complexes that is called *simple*. Plainly one has, by iterating (II), that τ is a simply homotopy invariant. In particular [C, (25.1)] τ is invariant under subdivision. This implies that for a smooth manifold, one can unambiguously define $\tau(E, D_i)$ to be the torsion of any smooth triangulation of K.

In case $K = S^1$, let A be the holonomy of a generator of $\pi_1 S^1$. One has that E is acyclic $\Leftrightarrow I - A$ is invertible $\Leftrightarrow$ E has no nonzero flat section and then

$$(III) \qquad \tau(E) = |\det I - A|$$

Note that the choice of generator is irrelevant as $I - A^{-1} = (-A^{-1})(I - A)$ and $|\det(-A^{-1})| = 1$.

These 3 properties of τ are the analogues of the properties of Euler characteristic noted above (cardinality law, homotopy invariance and normalization on a point), but there are differences. Since a point has no acyclic representations ($H^o \neq 0$) one cannot normalize τ on a point, and so one must use S^1 instead. The multiplicative cardinality law for τ can be made additive just by using $\log \tau$, so the difference here is inessential. More important for some purposes is that τ is *not* an invariant under a general homotopy equivalence: as mentioned earlier this is in fact why it was first invented.

It is not the case that the above 3 properties can be used to characterize τ, in the way that χ is determined by 3 axioms. Some further properties of τ describe its behavior under bundles.

Let $\pi: X \to B$ be a simplicial bundle with fiber F where F, B, X are finite complexes and π^{-1} sends subcomplexes of B to subcomplexes of X. Let E be a vector bundle over B and form its pullback $\pi^* E$ over X. We will give natural conditions on E under which $\pi^* E$ is acyclic and then compute $\tau(\pi^* E)$. Note that the vector spaces $H^i(\pi^{-1}p, \mathbf{C}), p \in B$, form a flat vector bundle over B, which we denote $H^i F$. The integral lattice in $H^i(\pi^{-1}p, \mathbf{R})$ determines a flat density by the condition that the covolume of the lattice be 1. We suppose that $E \otimes H^i F$ is acyclic for all i. Then $\pi^* E$ is acyclic and

$$(IV) \qquad \tau(\pi^* E) = \prod_i \tau(E \otimes H^i F)^{(-1)^i}$$

Note that in the above setting $\pi^* E$ is trivial on fibers. The opposite extreme is when one has a bundle E on X for which the restriction $E|F$ is acyclic. Then, for B connected,

$$(V) \qquad \tau(E) = \tau(E|F)^{\chi(B)}$$

In case F is finite, one can push forward a bundle E on X to bundle $\pi_* E$ on B. In terms of the holonomy representations, $\rho_{\pi_* E}$ is induced from the representation ρ_E of the subgroup $\pi_1 X \subset \pi_1 B$ (here we assume X, B connected). In general, the fiber of $\pi_* E$ at

$p \in B$ is the direct sum of the fibers E_q at $q \in \pi^{-1}p$. Clearly it inherits a flat density from that on E. Here one has that if E is acyclic so is $\pi_* E$ and

$$(VI) \qquad \tau(\pi_* E) = \tau(E)$$

This was proven in [NC].

These properties of τ of a bundle are related. If X is a product $X = X_1 \times X_2$ then one can fiber X with $B = X_1, F = X_2$. Then for an acyclic bundle E over X_1 and the pulled back bundle E' on X, one can compute $\tau(E')$ using (IV). Since $H^i F$ is trivial, one has $E \otimes H^i F = E \oplus \cdots \oplus E$ where one has $b_i = \dim H^i F$ summands. Thus $\tau(E \otimes H^i F) = \tau(E)^{b_i}$ and so (IV) gives $\tau(E') = \tau(E)^{\chi(B)}$. Taking $B = X_2, F = X_1$, one may apply (V) and get the same formula.

Also, if F is finite and E is a bundle over B with $E \otimes H_o F$ acyclic then (IV) follows from (VI). For $\pi_*(\pi^* E) = E \otimes H_o F$ and so $\tau(\pi^* E) = \tau(\pi_* \pi^* E) = \tau(E \otimes H_o F)$.

With these properties of τ behind us, we can see situations where τ counts circles. Suppose in (V) that $F = S^1$ i.e. X is a circle bundle. Then (V) can be regarded as saying that $\log \tau(E) = \chi(B) \log \tau(E|F)$ is counting the circle fibers in X in the way that χ counts points in B, with a weighting factor of $\log \tau(E|F)$.

A more important example is the dual situation of a bundle over S^1. Suppose in (IV) that $B \simeq S^1 = \mathbf{R}/\mathbf{Z}$. Choose a flow $\phi_t: X \to X$ so that for each $x \in X$ $\pi(\phi_t x)$ traces S^1 at unit speed in the positive direction (this can be done since X is locally a product of F and an interval). Then let $F_o = \pi^{-1}(0)$ and let $r = \phi_1 | F_o : F_o \to F_o$. Then r is a homeomorphism of the fiber F_o, determined by π up to an isotopy. It is called the *monodromy of* π (given the preferred generator g for $\pi_1 S^1$ implicit in our choice of homeomorphism $B \simeq \mathbf{R}/\mathbf{Z}$) or the *return map of* ϕ_t (for the given cross-section F_o). One can rephrase (IV) in terms of $r_i^*: H^i(F_o; \mathrm{C}) \to H^i(F_o; \mathrm{C})$ (beware that in some treatments it is this induced map that is called the monodromy of π). Let $A = \rho(g)$. Then the holonomy around g for $E \otimes H^i F$ is $A \otimes r_i^*$. So by (III) and (IV)

$$\tau(\pi^* E) = \prod_i |\det I - A \otimes r_i^*|^{(-1)^i}$$

Now consider the special case when E is one-dimensional, so A is just a complex scalar λ of modulus one, and we write E_λ for E. In terms of the rational function

$$\tilde{\varsigma}_r(t) = \prod_i \det(I - t r_i^*)^{(-1)^{i+1}}$$

we have

$$\tau(\pi^* E_\lambda) = |\tilde{\varsigma}_r(\lambda)^{-1}|$$

valid whenever the factors defining $\tilde{\varsigma}_r(t)$ do not vanish on λ, i.e. for all but a finite set of $\lambda \in S^1$.

But $\tilde{\varsigma}_r(t)$ is the *Lefschetz function* of r, that is the generating function for the sequence $L(r^n)$ of Lefschetz numbers of iterates of r:

$$\tilde{\varsigma}_r(t) = \exp \sum_{n>0} \frac{t^n}{n} L(r^n)$$

This identity follows formally from the formula

$$L(r^n) = \sum (-1)^i \operatorname{Trace}(r_i^*)^n$$

i.e. the homological formula for Lefschetz numbers. Using the geometric description of $L(r^n)$ as the intersection number of graph (r^n) and graph (id) in $F_o \times F_o$, we see that $\tilde{\varsigma}_r(t)$ is counting periodic points of r.

Now note that a point $x \in F_o$ fixed by r^n determines a closed orbit of ϕ of period n, namely $\phi_t x$, $0 \leq t \leq n$. This sets up a correspondence of periodic orbits of r and prime closed orbits of ϕ_t (where prime means that the circle in X is traced by the moving point exactly once, i.e. the orbit has multiplicity one). Thus one may equally regard $\tilde{\varsigma}_r(t)$ as counting the closed orbits of the flow ϕ. This means that the R-torsions $\tau(\pi^* E_\lambda)$ are counting circles once again, only this time the circles are transverse to the fibration.

For 2-dimensional E one can take A diagonal with eigenvalues λ, μ and compute $\tau(\pi^* E) = |\tilde{\varsigma}_r(\lambda)\tilde{\varsigma}_r(\mu)|^{-1}$. Letting λ, μ vary subject to $|\lambda\mu| = 1$ one recovers the divisor of $\tilde{\varsigma}_r$. Thus from these values of τ one can recover the Lefschetz numbers of the monodromy and its iterates.

In the special case $X = F \times S^1$, both discussions apply. We will see in §4 that fibrations over S^1 and oriented circle bundles fit into a general pattern in which R-torsion counts the closed orbits of a flow: the monodromy flow in the case of a fibration over S^1 and the flow along the fibers on an oriented S^1 bundle.

§2. Computations on chain complexes

In this section we will define R-torsion and prove the properties asserted for it in §1. We will also define R-torsion for a pair of finite complexes and prove corresponding properties.

Like Euler characteristic, the definition of R-torsion is based on algebra–only later does one give geometric applications. Roughly speaking Euler characteristic is a graded version of dimension, extending dimension from a single vector space to a complex of vector spaces. In a similar way, R-torsion is a graded version of the absolute value of the determinant of an isomorphism of vector spaces: for this to make sense, both vector spaces should be equipped with a positive density.

Recall that a *density* f on a complex vector space V of dimension n is a map $f: \Lambda^n V \to \mathbf{R}$ with $f(\lambda x) = |\lambda| f(x)$ for all $x \in V$, $\lambda \in \mathbf{C}$. The densities on V clearly form a real vector space $|V|$ of dimension one. If f is nonzero and takes values in $[0, \infty)$ we say it is *positive*. If V_1 and V_2 are both n-dimensional and $A: V_1 \to V_2$ is linear over $\mathbf{C}$ then there is an induced map $A^*: |V_2| \to |V_1|$. If each V_i carries a preferred positive density f_i then $A^* f_2 = a f_1$, for some $a \geq 0$, and we write $a = |\det A|$. In case $V_1 = V_2$, $f_1 = f_2$ then a is indeed the absolute value of the determinant of A, so this notation is consistent.

If $0 \to V_1 \to V_2 \to V_3 \to 0$ is an exact sequence of finite dimensional vector spaces over $\mathbf{C}$ then there is a natural isomorphism $\Lambda^{n_1} V_1 \otimes \Lambda^{n_3} V_3 \to \Lambda^{n_2} V_2$, $n_i = \dim V_i$. This induces a natural isomorphism $|V_1| \otimes |V_3| \simeq |V_2|$. If V is zero then absolute value is a standard generator for $|V|$ and sets up a natural isomorphism $|V| \simeq \mathbf{R}$: we will sometimes denote this generator for $|0|$ by 1.

Let $d^i: C^i \to C^{i+1}$ be a cochain complex of finite dimensional vector spaces over $\mathbf{C}$ with $C^i = 0$ for $i < 0$ or i large. Let $Z^i = \ker d^i$, $B^i = \operatorname{im} d^{i-1}$, and $H^i = Z^i / B^i$ be the cocycles, coboundaries, and cohomology of C^*. If one is given positive densities Δ_i on C^i and D_i on H^i (with the standard choice when i is negative or large) then R-torsion $\tau(C^*, \Delta_i, D_i) \in (0, \infty)$ is defined as follows.

The short exact sequences

$$0 \to Z^i \to C^i \xrightarrow{d^i} B^{i+1} \to 0$$

$$0 \to B^i \to Z^i \to H^i \to 0$$

together with the trivializations $|C^i| = \mathbf{R}\Delta_i$, $|H^i| = \mathbf{R} D_i$ give isomorphisms $|Z^i| \otimes |B^{i+1}| \simeq \mathbf{R}$, $|B^i| \simeq |Z^i|$ for each i. This gives isomorphisms $|B^i| \otimes |B^{i+1}| \simeq \mathbf{R}$, hence also $|B^i| \simeq$

$|B^{i+2}|$. But for $i \le 0$ or i large, $B^i = 0$ and so $|B^i| = \mathbf{R}$. So if j is large and even, we find $\mathbf{R} = |B^j| \simeq |B^0| = \mathbf{R}$ and this isomorphism is multiplication by a scalar $\tau \in \mathbf{R}^*$. Since positive densities are preserved under our isomorphisms, we find that $\tau \in (0, \infty)$. This is the *R-torsion* $\tau(C^*, \Delta_i, D_i)$. If $H^i = 0$ for all i we say C^* is *acyclic* and we write $\tau(C^*, \Delta_i)$ for the torsion when all the D_i are standard.

Suppose, for instance, that $C^o = C^1 = \mathbf{C}^n$ with the usual density $\Delta(e_1 \wedge \ldots \wedge e_n) = 1$ and $C^i = 0$ for other i and $H^i = 0$ for all i. Then $1 \in |B^2|$ corresponds to the standard generator Δ for $|B^1| = |\mathbf{C}^n|$, which in turn corresponds to $|\det d^o|^{-1} \in |Z_o|$ and to $|\det d^o| \in |B^o|$. Thus $\tau = |\det d^o|$, showing that τ generalizes the absolute value of the determinant.

Suppose that $C_i = 0$ for $i > 1$. Then d^o induces an isomorphism $A\colon C^o/Z^o \to B^o$. Each of these spaces is equipped with a natural density, namely Δ_o/D_o and Δ_1/D_1 in an obvious notation. In this case $\tau = |\det A|$, i.e. $A^*(\Delta_1/D_1) = \tau(\Delta_o/D_o)$. So here τ is the absolute value of the determinant of the nonsingular part of d^o.

It is easy to see how the choice of densities affects τ:

$$\tau(C^*, \alpha_i\Delta_i, \beta_i D_i) = \prod_j (\beta_j/\alpha_j)^{(-1)^j} \cdot \tau(C^*, \Delta_i, D_i)$$

where $\alpha_i, \beta_i > 0$ and where we take $\alpha_i = \beta_i = 1$ if i is negative or large.

A case of special interest is when H^i, C^i have a preferred basis for each i. This defines preferred generators D^i, Δ^i and so τ is defined. This agrees then with the Milnor definition of R-torsion [M3,C] which may be regarded as a computational device.

When C is acyclic, there is another way to compute τ. This is to take a chain contraction $\delta^i\colon C^i \to C^{i-1}$, i.e. a linear map with $d\delta + \delta d = id$. Then $d + \delta$ determines a map $(d+\delta)_+\colon C^+ = \oplus C^{2i} \longrightarrow C^- = \oplus C^{2i+1}$ and a map $(d+\delta)_-\colon C^- \to C^+$. As $(d+\delta)^2 = id + \delta^2$ is unipotent, $(d+\delta)_+$ is an isomorphism. One can show $\tau = |\det(d+\delta)_+|$ [C].

If one has a short exact sequence of such complexes $0 \to C'^i \to C^i \to C'''^i \to 0$ and (with obvious notation) the densities Δ_i' and Δ_i'' determine Δ_i then one can compare the R-torsions of C', C, and C''. Namely let $0 \to H^oC' \to H^oC \to H^oC'' \to H^1C' \to \ldots$ be the usual long exact sequence regarded as a cochain complex $\hat{C}$ with $\hat{C}^{3i} = C'^i, \hat{C}^{3i+1} = C^i$ and $\hat{C}^{3i+2} = C'''^i$. If one gives $\hat{C}^*$ the densities $\hat{D}_{3i} = D_i', \hat{D}_{3i+1} = D_i$ and $\hat{D}_{3i+1} = D_i''$, then we have

$$(*)\qquad \tau(C,\Delta_i,D_i)=\tau(C',\Delta_i',D_i')\tau(C'',\Delta_i'',D_i'')\tau(\hat{C},\hat{D})$$

This was proven (in terms of preferred bases and for homology instead of cohomology) by Milnor [M3, C]. In particular, when C, C', C'' are acyclic (it is enough to assume that 2 are and then the 3rd must be)

$$(**)\qquad \tau(C,\Delta_i)=\tau(C',\Delta_i')\tau(C'',\Delta_i'')$$

From $(*)$ one may easily deduce the behavior of $\tau(C)$ when Δ_i is not determined by Δ_i', Δ_i''. One finds, by our formula for charge of densities,

$$(*)'\qquad \tau(C,\Delta_i,D_i)=\tau(C'^i,\Delta_i',D_i')\tau(C''^i,\Delta_i'',D_i'')\tau(\hat{C}^i,\hat{D}_i)(\tau_-/\tau_+)$$

where τ_+ (respectively τ_-) is the torsion of the exact sequence

$$0\to\oplus_i C'^i\to\oplus_i C^i\to\oplus_i C''^i\to 0$$

viewed as a cochain complex with terms in degrees 0, 1 and 2 and where i varies over even (respectively odd) integers.

An important special case of $(*)$ is when C is the direct sum of C' and C'', and one has $D_i = D_i'D_i'', \Delta_i = \Delta_i'\Delta_i''$. Then $\tau(C)=\tau(C')\tau(C'')$, i.e. τ is multiplicative over direct sums.

Now we turn to the geometric application. Suppose E is a flat bundle over a finite complex K that admits a flat density, i.e. $|\det \rho_E| = 1$. Let V denote the fiber. Then the cochain complex $C^i(K;E)$ can be identified with a direct sum of a copy of V for each i-cell σ in K, by choosing a basepoint in each component of K and a basepath from each i cell to the basepoint. This gives an isomorphism $|C^i(K;E)| = \otimes_\sigma |V|$. Because $|\det \rho_E| = 1$, this isomorphism is independent of the choice of basepaths. Choosing a flat density on E, we obtain a preferred density $\Delta_i \in |C^i(K;E)|$. Given a density D_i on each $H^i(K;E)$, one defines the R-torsion of (E, D_i)

$$\tau(E,D_i)=\tau(C^*(K;E),\Delta_i,D_i)\in(0,\infty)$$

One can compute the Euler characteristic of $C^i(K;E)$ on the chain level, obtaining $(\dim V)\chi(K)$, or on the cohomology level. Thus if E is acyclic then each component K_0 of

K has zero Euler characteristic (as $E|K_o$ is also acyclic). If the flat density in each fiber is multiplied on K_0 by a factor $\lambda > 0$ then the value of τ changes by a factor $\lambda^{(\dim V)\chi(K_0)}$. So if E is acyclic, one sees that $\tau(E)$ is independent of the choice of flat density, as asserted in §1.

Now suppose A and B are subcomplexes of K. Then there is a twisted Mayer-Vietoris short exact sequence

$$0 \to C^i(A \cup B, E|A \cup B) \to C^i(A, E|A) \oplus C^i(B, E|B) \to C^i(A \cap B, E|A \cap B) \to 0$$

and a corresponding long exact sequence. This shows that if three of $E|A, E|B, E|A \cup B, E|A \cap B$ are acyclic then so is the fourth. Noting that the conditions on the $\Delta_i, \Delta_i', \Delta_i''$ are satisfied, we see that the cardinality law (I) of Section 1 follows from formula (**) above.

We now prove (II) of Section 1. Define

$$C^i = C^i(K; E), C''^i = C^i(K'; E|K') \text{ and } C'^i = \ker(C^i \to C''^i)$$

Then the cohomology long exact sequence shows that C'^i is acyclic. Evaluating cochains on σ, σ', we identify C'^i with V for $i = n, n-1$. As $C'^i = 0$ for other i the conditions on $\Delta_i, \Delta_i', \Delta_i''$ hold and (**) applies. But the boundary map in C'^i is just ± 1 and so $\tau(C') = 1$. This gives (II).

The proof of (III) is simple. One chooses the cell complex on S^1 with one 0-cell and one 1-cell. Then $C^i(K; E) = V$, $i = 0, 1$, and vanishes for other i. The differential $d^o: C^o \to C^1$ is $\pm(I - A)$, where A is the holonomy of a generator of $\pi_1 S^1$, depending on the various choices of orientation. So $\tau(E) = |\det I - A|$.

We now prove (V). If B is a simplex then $X = B \times F$ has the simple homotopy type of F. So using (II) and $\chi(B) = 1$ gives (V). Now induct on the number of cells in B. Suppose B is obtained by attaching a cell σ of top dimension to a subcomplex B_0. Then using the cardinality law (II) for $\pi^{-1}B_0$, $\pi^{-1}\overline{\sigma}$ gives

$$\begin{aligned}\tau(E) &= \tau(E|\pi^{-1}B_0)\tau(E|\pi^{-1}\overline{\sigma})/\tau(E|\pi^{-1}(B_0 \cap \overline{\sigma})) \\ &= \tau(E|F)^{\chi(B_0)+\chi(\overline{\sigma})-\chi(B_0\cap\overline{\sigma})}\end{aligned}$$

by induction and our result for simplices. Now use the cardinality law for χ to obtain (V).

Before proving (IV) in general, we sketch a proof for the important case $B = S^1, E = E_\lambda$. Regard B as a cell complex with one vertex v and one edge. Then with $F = \pi^{-1}v$, the exact sequence $\hat{C}$ of the pair (X,F) with coefficients in $\pi^* E_\lambda$ can be written as a Wang sequence

$$\to H^i X \to H^i F \xrightarrow{\partial} H^i F \to H^{i+1} X \to$$

where $\partial = I - \lambda r_i^*$. Thus $\pi^* E_\lambda$ is acyclic iff $\det(I-\lambda r_i^*) \neq 0$, all i, and $\tau(\hat{C}) = |\tilde{\varsigma}_r(\lambda)^{-1}|$. Using $(*)$, with $C = C^*(X;\pi^* E)$ and $C'' = C^*(F;\pi^* E|F)$ one finds $\tau(C) = \tau(C')\tau(C'')\tau(\hat{C})$. But C' is a relative cochain complex for $(F \times [0,1], F \times \partial[0,1])$ and so has R-torsion the reciprocal of $\tau(C'')$ (the complexes differ by 1 in degree). Thus $\tau(\pi^* E_\lambda) = \tau(C) = \tau(\hat{C}) = |\varsigma_r(\lambda)^{-1}|$. Another proof of this case, using the chain contraction definition of τ and special properties of a mapping torus, is given in [F1]: we use this approach in §3.

Our proof of (IV) is based on the Serre spectral sequence with coefficients in $\pi^* E$. The hypothesis that $H^i F \otimes E$ is acyclic means that the E_2 term $H^*(B; H^* F \otimes E)$ of this spectral sequence vanishes, so $\pi^* E$ is acyclic. We then need to study the algebraic behavior of τ in such a spectral sequence.

Let C^i be a cochain complex over $\mathbf{C}$ as before but equipped with a filtration $\subset F_n^i \subset \ldots \subset F_0^i = C^i$ such that $F_{i+1}^i = 0$ (such a filtration is called *canonically cobounded*). Suppose that each $F_p^i, H^i(F_p)$, and $H^i(F_p/F_{p-1})$ has a preferred positive density, Δ_{pi}, D_{pi}, and δ_{pi} respectively, with our conventional choice for the zero vector space. We give F_p^i/F_{p-1}^i the quotient density $\Delta_{pi}/\Delta_{p-1,i}$. Let $E_1^{pq} = H^{p+q}(F_p/F_{p+1})$ with the coboundary $d_1: E_1^{pq} \to E_1^{p+1,q}$ obtained as the coboundary for the short exact sequence $0 \to F_{p+1}/F_{p+2} \to F_p/F_{p+2} \to F_p/F_{p+1} \to 0$. We suppose $E_2 = 0$, so d_1 is exact for each fixed q. Let the *associated graded complex* be $GC^i = \oplus_p F_p^i/F_{p+1}^i$. Then each GC^i and $H^i(GC)$ has a preferred positive density and so we have an R-torsion that we denote $\tau(GC)$ (here and for the rest of the proof we omit mention of densities for notational convenience). We will find that this is related to $\tau(C)$ with an error coming from the exact sequences d_1^{*q}:

$$(***) \qquad \tau(C) = \tau(GC) \prod_q \tau(E_1^{*q})^{(-1)^q}$$

Given this, we take B^p to be the p-skeleton of B, with the convention $B^{-1} = \emptyset$, and set $F_p^i = \ker(C^i(X) \to C^i(\pi^{-1} B^{p-1})) \subset C^i(X)$, where the cochains have coefficients in $\pi^* E$. Then this is the Serre spectral sequence for $\pi^* E$. We choose the Δ_{ki} as in the definition of R-torsion of a complex. Here the associated graded complex is just the direct sum over each closed cell σ in B of the cochains for the pair $(\pi^{-1}\sigma, \pi^{-1}\partial\sigma)$, which is the same as cochains for $F \times (D, \partial D)$ where D is a ball of dimension $\dim \sigma$. Thus $\tau(GC) = \prod_\sigma \tau(F)^{(-1)^{\dim \sigma}} = \tau(F)^{\chi(B)}$. But since B has an acyclic bundle ($E \otimes H^0 F$, say) we have $\chi(B) = 0$, so $\tau(GC) = 1$. On the other hand d_1^{*q} is the coboundary operator for $C^*(X, E \otimes HF)$ and so (IV) follows from $(***)$.

To prove $(***)$, we first note that $E_2 = 0$ implies $H^i F_{p+1} \to H^i F_p$ is zero for all i, p. This is easily seen by setting up the exact couple of the filtration: $A = \bigoplus_{p \in \mathbf{Z}} H(F_p)$, $B = \bigoplus_{p \in \mathbf{Z}} H(F_p / F_{p+1})$ where $F_p = F_0$ for $p \leq 0$ with maps $h: A \to A$, $j: A \to B$, $k: B \to A$ defined as the direct sum over p of the long exact sequences of the pairs (F_p, F_{p+1}). Then $0 = E_2 = H(B)$ so the derived exact couple has $B' = 0$ and $h': A' \to A'$ an isomorphism. Now $A' = h(A)$, h' shifts the terms in the direct sum

$$h(A) = \bigoplus_{p \in \mathbf{Z}} im\ (H(F_{p+1}) \to H(F_p)),$$

and these terms are zero for p large. Thus all the terms are zero, so $h = 0$.

Hence the long exact sequence $\hat{F}_p$ of (F_p, F_{p+1}) gives for each i a short exact sequence (trivial unless $0 \leq i \leq p$)

$$(S_{ip}) \qquad 0 \to H^i F_p \xrightarrow{j} H^i(F_p / F_{p+1}) \xrightarrow{k} H^{i+1} F_{p+1} \to 0$$

One can view this as a cochain complex with terms of degree 0, 1 and 2. Then one has

$$\tau(\hat{F}_p) = \prod_{i \geq 0} \tau(S_{ip})^{(-1)^{i+1}}$$

On the other hand, we can view (S_{ip}) as a short exact sequence of chain complexes

$$0 \to C'_q \xrightarrow{j} C_q \xrightarrow{k} C''_q \to 0$$

with $C'^i_q = H^i F_p$, $C^i_q = H^i(F_p / F_{p+1})$ and $C''^i_q = H^{i+1} F_{p+1}$ for $i = p + q$. Here the coboundary operators on C'_q, C''_q are zero whereas that on C_q is $d_1 = jk$. We now apply $(*)'$ to C_q. As C'_q, C''_q only differ by their grading, we have $\tau(C'_q)\tau(C''_q) = 1$. As the long exact sequence $\hat{C}_q$ consists of identity isomorphisms $C''^i_q \to C'^{i+1}_q$, $\tau(\hat{C}_q) = 1$. One find then that $(*)'$ is

$$\tau(C_q) = \prod_{i \geq 0} \tau(S_{i,i-q})^{(-1)^{i+1}}$$

Now we compute

$$\begin{aligned}\tau(C) &= \prod_{p\geq 0} \tau(F_p)/\tau(F_{p+1}) \\ &= \prod_{p\geq 0} \tau(F_p/F_{p+1})\tau(\hat{F}_p) \\ &= \tau(GC) \prod_{p,i\geq 0} \tau(S_{ip})^{(-1)^{i+1}} \\ &= \tau(GC) \prod_{q\geq 0} \tau(C_q)\end{aligned}$$

But since $E_1^{pq} = C_q^{p+q}$ we have $\tau(C_q) = \tau(E_1^{*q})^{(-1)^q}$ (there is a degree shift of q) proving $(***)$ and hence (IV).

We now turn to the corresponding results for pairs. Suppose L is a CW subcomplex of K and form the relative cochain complex $C^*(K,L;E) = \ker\,(C^*(K;E) \to C^*(L;E|L))$. Then one has a natural isomorphism $|C^*(K,L;E)| \simeq \otimes_j |V|$ where j runs over the i-cells in $K - L$. So our flat density on E gives a density Δ_i on the relative i-cochains in E. Thus we again have an R-torsion, for some choice of positive densities D_i on $H^i(K,L;E)$, that we denote $\tau(K,L;E,D_i)$. If $H^i(K,L;E) = 0$ we say E is acyclic for (K,L) and we simply write $\tau(K,L;E)$, when the D_i are chosen standard. Since we now are writing out K and L explicitly, we will no longer be so explicit about restricting flat bundles to subcomplexes: this should not cause confusion.

There are more or less obvious analogues of (I)-(VI) for pairs. If L_i is a subcomplex of K and M_i a subcomplex of L_i, $i = 1,2$, then

$$(I) \qquad \tau(L_1 \cup L_2, M_1 \cup M_2; E) = \tau(L_1, M_1; E)\tau(L_2, M_2; E)\tau(L_1 \cap L_2, M_1 \cap M_2; E)$$

where we assume that E is acyclic for 3 of the 4 pairs in the formula. The proof goes as before. Also if $C \subset B \subset A \subset K$ are subcomplexes, one has

$$(I)' \qquad \tau(A,B;E)\tau(B,C;E) = \tau(A,C;E)$$

where one assumes E is acyclic for 2 of the 3 pairs involved. The proofs use $(**)$ on the Mayer-Vietoris short exact sequence for pairs and the short exact sequence of a triple, respectively.

A good notion of simple homotopy for pairs should allow for excision as well as for elementary collapses. Suppose K', σ, σ' are as in Section 1, and $L \subset K'$ is a subcomplex. Let (K'', L'') be a pair of subcomplexes of K with $K'' - L'' = K - L$. Then

$$(II) \qquad \tau(K', L; E) = \tau(K, L; E) = \tau(K''. L''; E)$$

Here the first equality goes as before and the second is trivial, since the cochain complexes for (K, L) and (K'', L'') are identical. If we define a simple homotopy of pairs to be any sequence of such elementary collapses and excisions, as well as their inverses, then we see that R-torsion of a pair is invariant under simple homotopy provided that the flat bundle can be extended through the inverse excisions. This is not always possible, as one sees if $K = S^2 \vee D^2$, $L = D^2$, $K'' = S^2 \vee S^1$, $L'' = S^1$ and E'' is a flat bundle on K'' that is not a product on L''.

Now take $\pi: X \to B$ as before. Suppose X' is a subcomplex of X such that $\pi|X'$ is also a fiber bundle with fiber $F' \subset F$. Then

(IV) If E is a flat bundle on B with a flat density such that $E \otimes H^i(F, F')$ is acyclic for all i then $\pi^* E$ is acyclic for (X, X') and

$$\tau(X, X'; \pi^* E) = \prod_i \tau(B; E \otimes H^i(F, F'))^{(-1)^i}$$

Now let $B' \subset B$ be a subcomplex, set $X' = \pi^{-1} B' \subset B$ and let E be a bundle on X with a flat density.

(V) If E is acyclic for F then E is acyclic for (X, X') and

$$\tau(X, X'; E) = \tau(F; E)^{\chi(B, B')}$$

(VI) If F is finite and E is acyclic for (X, X') then $\pi_* E$ is acyclic for (B, B') and

$$\tau(B, B'; \pi_* E) = \tau(X, X'; E)$$

The proofs of (IV) - (VI) are just like the cases treated above, using the relative cochain complexes instead of absolute ones.

Suppose $B = S^1$ and $F = D^u$ is a disc of dimension u. Let $\Delta(E) = 1$ if π is an oriented disc bundle, -1 if it isn't. Let $\epsilon(E) = (-1)^u$ and let $A = \rho_E(g), g \in \pi_1 S^1$ a generator.

(III) Provided the determinant is nonzero

$$\tau(X, \partial X; \pi^* E) = |\det I - \Delta(E) A|^{\epsilon(E)}$$

This follows from (IV) above with $X' = \partial X$. Setting $u = 0$ gives (III) of §1.

Note that for each n, the set of representations $\rho: \pi_1 K \to Gl(n, \mathbf{C})$ is an algebraic set and that for each i the set S_i of ρ for which $H^j(K, L; E_\rho) = 0$ for $0 \le j \le i$ is Zariski open in S_{i-1}, being defined by the condition that the coboundary d^i have minimal possible nullity. Thus the acyclic ρ's are Zariski open (let $i = \dim K$). Then one sees easily that $\tau(\rho)$ is a regular function on the real affine algebraic set of acyclic ρ with $|\det \rho| = 1$. Specializing to $K = X, L = \partial X$ as in the preceding paragraph, we parametrize ρ by $A \in Gl(n, \mathbf{C})$, $n = \dim E$, by choosing a basis on a fiber of E. From the continuity of $\tau = \tau(A)$, it follows that the A's where the determinant vanishes cannot be acyclic. This can also be seen from the Wang sequence for the pair $(X, \partial X)$.

§3 Smooth flows with circular R

We now apply the topological tools of the preceding sections to smooth flows. Using terminology that we review below, the key steps are

(1) an isolated invariant set I determines a simplicial pair up to simple homotopy, and hence one can attach to it certain R-torsions.

(2) when I is circular these R-torsions can be computed from closed orbits.

(3) the chain recurrent set R can be approximated by isolated invariant sets.

Altogether, when R is circular we will express the R-torsion of certain flat bundles in terms of closed orbits, generalizing the results of §1 for the bundles $\pi^* E_\lambda$.

For example, suppose ϕ is a nonsingular almost Morse-Smale (or NAMS) flow ϕ on a closed manifold M. Here NAMS means ϕ has finitely many hyperbolic prime periodic orbits γ and no other chain recurrent points. Over γ lies a strong unstable bundle $E^u(\gamma)$ of some dimension $u(\gamma)$. Let $\Delta(\gamma)$ be $+1$ if E^u is orientable, -1 if it is not. Let $\epsilon(\gamma) = (-1)^{u(\gamma)}$. We will show that for a flat bundle E over M with a flat density:

Theorem 3.1 $\tau(E) = \prod_\gamma |\det I - \Delta(\gamma)\rho_E(\gamma)|^{\epsilon(\gamma)}$ *provided that each determinant occurring in this formula is nonzero (in particular, this condition on E assures that E is acyclic.)*

We now define R. Suppose X is a smooth compact manifold and $\phi_t: X \to X$ the local flow associated to a smooth vector field transverse to ∂X. Let $\partial_- X \subset \partial X$ be the outset of ϕ. A (T, ϵ)-*chain* is a path consisting alternately of arbitrary short paths of length $\le \epsilon$

and long segments of flowline of the form $\phi_t y, 0 \le t \le t_0$, $t_0 \ge T$. Here length is measured by some Riemannian metric on X. If for all $T, \epsilon > 0$ there is a (T,ϵ)-chain beginning and ending at x, we say x is *chain-recurrent* for ϕ. The set $R = R(\phi)$ of such x is the *chain-recurrent set* of ϕ. Clearly R is invariant (i.e. closed in X and $x \in R \Rightarrow \phi_t x \in R$ for all $t \in \mathbf{R}$) and contains all the closed orbits.

If $\mathcal{O} \subset X$ is such that $\bigcap_{t \in \mathbf{R}} \phi_t \mathcal{O} = I$ is compact and $I \subset int\mathcal{O}$ then we say $\mathcal{O}$ is an isolating neighborhood for the isolated invariant set I. Any neighborhood $\mathcal{O}'$ of I with $\mathcal{O}' \subset \mathcal{O}$ is then also an isolating neighborhood for I. It follows that ϕ admits at most a countable number of isolated invariant sets (X has a countable basis). They are the most tractable invariant sets of ϕ.

Now we recall a result of Conley [Co]: there is a continuous function $G: X \to \mathbf{R}$ such that G is strictly decreasing on $X - R$ and GR is nowhere dense in $\mathbf{R}$. Taking U an open neighborhood of GR and $\mathcal{O} = G^{-1}U$ we see that $\mathcal{O}$ is an isolating neighborhood for some I and that $I \to R$ as $U \to GR$. This proves step (3).

For step (1) we need the isolating blocks of Conley-Easton and Montgomery [CE,Mo]. Given an isolated invariant set I, an arbitrarily small compact isolating neighborhood B for I can be chosen with the following properties

(a) B is a smooth manifold with corners

(b) $\partial B = b^+ \cup b^- \cup b^o$ where each term is a compact manifold with boundary

(c) ϕ is tangent to b^o and makes b^o a smooth interval bundle with $\partial b^o = b^o \cap (b^+ \cup b^-)$

(d) ϕ is transverse to b^+, b^-, enters B on b^+ and exits on b^-.

Such a B we will call an *M-block* after Montgomery–it is a variation on the earlier construction of Conley and Easton. The pair (B, b^-) is the simplicial pair we use.

There are 2 elementary moves on M-blocks called squeezing and shaving. A *squeeze* B_1 of B is determined by compressing B along the flow: given smooth functions $\tau_+ : b^+ \to [0,\infty)$ and $\tau_- : b^- \to [0,\infty)$ such that $\phi_t x \ne \phi_s y$ for $0 \le t \le \tau_+(x)$, $-\tau_-(y) \le s \le 0$, $x \in b^+, y \in b^-$

$$B_1 = B - \{\phi_t x | x \in b^+,\ 0 \le t < \tau_+(x)\} - \{\phi_s y | y \in b^-, -\tau_-(y) < s \le 0\}$$

A *shave* B_0 of B is the subset obtained by deleting an open set of trajectories joining b^+ to b^-: if $b_0^+ \subset b^+$, $b_0^- \subset b^-$ are compact submanifolds with boundary (of dimension

$\dim X - 1)$ such that $U = \overline{b^+ - b_0^+}$ and $V = \overline{b^- - b_0^-}$ are also submanifolds with boundary and such that for each $x \in U$ there is an $\tau(x) > 0$ with $\phi_t x \in B$, $0 \le t \le \tau(x)$, and $\phi_{\tau(x)}(x): U \to V$ a diffeomorphism,

$$B_0 = B - \{\phi_t x | x \in int(U),\ 0 \le t \le \tau(x)\}.$$

These are minor modifications of Montgomery's definitions [Mo] to the smooth category.

If B_0 is a shave of B then there is a sequence of elementary collapses from B to $B_0 \cup b^-$ and an excision from $(B_0 \cup b^-, b^-)$ to (B_0, b_0^-). Thus the simple homotopy type of (B, b^-) isn't changed by a shave. It is clear that the diffeomorphism type of (B, b^-) isn't changed by a squeeze. But by [Mo] (with obvious modifications to account for smoothness) any two M-blocks for a given isolated I are related by a sequence of shaves and squeezes. This establishes step (1). In an obvious terminology, one may say that shave/squeeze equivalence of M-blocks implies simple homotopy equivalence of the pairs (block, exit set).

A basic example is when I is a hyperbolic prime closed orbit γ. Then one can choose B to be a bundle over S^1 with fiber F a simplicial disc so that $F \simeq I^u \times I^s$, $u = u(\gamma)$, $s = s(\gamma)$ and $F \cap b^- \simeq \partial I^u \times I^s \subset F$. By collapsing I^s to a point one can produce a simple homotopy equivalence of (B, b^-) to $(X_\gamma, \partial X_\gamma)$ where X_γ is the unstable disc bundle over γ.

We can now prove Theorem 1. Give M its Smale filtration by compact submanifolds M_i of top dimension so that $M_i \subset$ int M_{i+1}, $M_0 = \emptyset, M_i = M$ for i large, ϕ is transverse inward on ∂M_i, and $M_{i+1} - M_i$ is an isolating neighborhood for a hyperbolic closed orbit γ. Then (M_{i+1}, M_i) has the same simple homotopy type as $(X_\gamma, \partial X_\gamma)$. One finds

$$\begin{aligned}\tau(E) &= \prod_i \tau(M_{i+1}, M_i; E) \\ &= \prod_\gamma \tau(X_\gamma, \partial X_\gamma; E) \\ &= \prod_\gamma |\det I - \Delta(\lambda)\rho_E(\gamma)|^{\epsilon(\gamma)}\end{aligned}$$

by (III) of section 2. Reading the equations from bottom to top, one justifies the acyclicity of E over $(X_\gamma, \partial X_\gamma)$, then for (M_{i+1}, M_i) and finally for M.

Now for step (2). An invariant set I for ϕ is *circular* if there is a smooth map $\theta: U \to$ $\mathbf{R}/\mathbf{Z}$, U a neighborhood of I in X, on which $\frac{d}{dt}(\theta \circ \phi_t) > 0$. In other words, there is a cross-section to ϕ on I, namely a level set of $\theta|\text{int}(U)$. Circular invariant sets abound.

For instance if I is finite (i.e. consists of finitely many closed orbits) then I is circular. More generally if $\phi|I$ has no stationary points and the topological dimension of I is 1 then I is circular. With some work it can be shown that circularity does not depend on the embedding, i.e. I is circular for ϕ $\leftrightarrow$ there is a fibration of I over S^1 transverse to $\phi|I$. For criteria for when a given I is circular see [Sc,Fu,F2].

Now suppose I is circular and isolated. Then we can take U to be an M-block B for I, since arbitrarily small M-blocks exist. By perturbing θ, we can make θ transverse to $0 \in \mathbf{R}/\mathbf{Z}$ on $B, b^+, b^-, b^o, \partial b^+, \partial b^-$ and ∂b^o. Now let $Y = \theta^{-1}(0) \cup b^-$, $Z = b^-$. Then (Y, Z) is a simplicial pair.

We define a continuous map $r: Y \to Y$ as follows. If $y \notin Z$ then $r(y) = \phi_\tau y$ where $\tau = \tau(y) > 0$ is the smallest positive time t for which $\phi_t y \in Y$. Since ϕ is transverse out on $b_- = Z$, we see that $\tau(y)$ is near 0 for y near Z: thus τ extends continuously to Y if we set $\tau|Z = 0$. Now let $r(y) = \phi_{\tau(y)} y \in Y$. Note that $r^{-1}(Z)$ is a neighborhood of Z in Y, by this same transversality.

Now let E be a flat bundle over B. There is a bundle map $\alpha: r^*(E|Y) \to (E|Y)$ defined by pulling back along the trajectory from y to $r(y)$, using the flat connection on E. This determines an endomorphism $\alpha_*: H^*(Y, Z; r^*E) \to H^*(Y, Z; E)$. Since there is a natural induced map $r^*: H^*(Y, Z; E) \to H^*(Y, Z; r^*E)$, we obtain the endomorphism $\beta = \alpha_* r^*$ of $H^*(Y, Z; E)$.

Suppose, more generally, that (Y, Z) is a simplicial pair $f: (Y, Z) \to (Y, Z)$ a map, E a flat bundle over Y and $\alpha: f^*E \to E$ a locally constant bundle map. Then $\alpha_* f^*$ is an endomorphism of $H^*(Y, Z; E)$ and so the relative Lefschetz number $L = L(\alpha_* f^*) = \sum_i (-1)^i$ Trace $(\alpha_i f^{*i})$ is defined. If $f^{-1}Z$ is a neighborhood of Z in Y then L can be computed from the fixed point set of f in $Y - Z$. Indeed, if Fix $f - Z$ is a finite set of points p with Lefschetz index $\mathrm{ind}_L(f, p)$ and $\alpha_p: E_p \to E_p$ is the endomorphism of the fiber at p then one has the relative Lefschetz formula

$$(RL) \qquad L = \sum_p \mathrm{ind}_L(f, p) \cdot \text{ Trace } \alpha_p$$

This is proven analytically by Atiyah and Bott when $Z = \emptyset$, f smooth and all p are simple [AB] but it is really a purely topological result; the proof of the Lefschetz formula (for $Z = \emptyset$, E trivial) in Franks' notes [Fr2] can be readily adapted to this level of generality. One needs $r^{-1}Z$ to be a neighborhood of Z in order that fixed points in Z count equally towards the Lefschetz numbers of $f: Y \to Y$ and $f|Z: Z \to Z$ so that they contribute zero to the relative Lefschetz number. This remark allows one to reduce (RL) to the absolute case $(Z = \emptyset)$.

Returning to the case at hand, we see that $L(\beta^k)$, $k \geq 1$, counts the points of period k for r not in Z, i.e. the closed orbits of ϕ that wrap k times around $\mathbf{R}/\mathbf{Z}$ under θ, with a weight coming from the holonomy of E around these closed orbits. Thus the exponential generating function

$$\varsigma_\beta(t) = \exp \sum_{k>0} L(\beta^k) t^k / k$$

can be computed from the closed orbits in I. We call it the *twisted Lefschetz function* for E and (B, b^-).

Suppose, for instance, that I is a prime hyperbolic closed orbit γ and that $\theta|\gamma$ has degree 1. Let $p = \gamma \cap Y$. Then $\alpha_p = \rho_E(\gamma)^{-1}$ so

$$L(\beta^k) = \epsilon(\gamma)\Delta(\gamma)^k \text{ Trace } \rho_E(\gamma)^{-k}$$

where we have used the formula for the index of a hyperbolic fixed point. Using the formal power series identity

$$(D) \qquad \det I - tA = \exp - \sum_{k>0} \text{Trace } A^k \cdot t^k / k$$

for $A = \Delta(\gamma)\rho_E(\gamma)^{-1}$ one finds

$$\varsigma_\beta(t) = \det(I - \Delta(\gamma)\rho_E(\gamma)^{-1} t)^{\epsilon(\gamma)+1}$$

Setting $t = 1$, we see $|\varsigma_\beta(1)|^{-1}$ is the contribution of γ to $\tau(E)$ in Theorem 1.

We now generalize this to other circular isolated I corresponding to step (2) above.

Theorem 3.2 *E is acyclic for (B, b^-) $\Leftrightarrow$ $I - \beta$ is invertible and then*

$$\tau(B, b^-; E) = |\varsigma_\beta(1)|^{-1}$$

We remark that if X is circular (i.e. ϕ has a global cross-section) then Theorem 3.2 shows that the computation of §1 for $\pi^* E_\lambda$ holds for any E acyclic for $(X, \partial_- X)$.

We map $Y \times [0,1]$ to B by sending (y, t) to $\phi_{t\tau(y)} y$. Then $(y, 1)$ and $(ry, 0)$ both map to ry, so one has an induced map $\sigma: M \to B$, where $M = M_r$ is the mapping torus of r.

We let $N = Z \times S^1$ be the subset swept out by Z. Clearly $N = \sigma^{-1}(b^-)$. If θ is chosen generically then (M, N) is a simplicial pair and σ is PL. Then one has an evident simple homotopy equivalence from (B, b^-) to $(\sigma(M), b^-)$ by pushing along the flow. The PL map $\sigma: (M, N) \to (\sigma(M), b^-)$ is a relative homeomorphism. Thus

$$H^*(M, N; \sigma^* E) = H^*(B, b^-; E)$$

and if this vanishes

$$\tau(M, N; \sigma^* E) = \tau(B, b^-; E)$$

Now we homotope r to a simplicial map $\hat{r}: Y \to Y$ that fixes Z and we let $\hat{M}$ be the mapping torus of $\hat{r}$. Let $\hat{E}$ be the flat bundle over $\hat{M}$ corresponding to $\sigma^* E$. Again $(\hat{M}, N)$ has the same cohomology and torsion as (M, N) c.f. [Co](5.5). One can give $(\hat{M}, N)$ a CW complex structure for which each cell σ in Y corresponds to two cells in $\hat{M}$, σ itself and σ' = the image of $\sigma \times [0,1]$ in $\hat{M}$. One finds that this CW complex has a boundary operator with block form

$$\begin{pmatrix} \partial_Y & s \\ 0 & -\partial_Y \end{pmatrix}$$

where the first column corresponds to $C_* Y \subset C_* \hat{M}$, with ∂_Y the boundary operator on ∂_Y, and where the second column corresponds to $C_* \hat{M} / C_* Y \simeq C_{*-1} Y$ via the correspondence $\sigma' \to \sigma$. Here $s(\sigma') = \hat{r}_* \sigma - \sigma$.

Now let E_λ be the flat line bundle on S^1 with holonomy $\lambda \in \mathbf{C}^*$ and let $\pi: \hat{M} \to S^1 = [0,1]_{/0=1}$ be the natural map. We then have for the bundle $\hat{E} \otimes \pi^* E_\lambda$ and the pair $(\hat{M}, N)$ a coboundary operator

$$d = \begin{pmatrix} d_E & 0 \\ s_\lambda & -d_E \end{pmatrix}$$

where d_E is the coboundary for E for the pair (Y, Z) and where $s_\lambda = \lambda \hat{r}^* - id$. For almost all $\lambda \in S^1$, s_λ is invertible. Then

$$\delta = \begin{pmatrix} 0 & s_\lambda^{-1} \\ 0 & 0 \end{pmatrix}$$

satisfies $d\delta + \delta d = I$, since s_λ commutes with d_E. Hence $\hat{E} \otimes \pi^* E_\lambda$ is acyclic for $(\hat{M}, N)$ with torsion $\tau_\lambda = |\det(d + \delta)_+|$. But

$$(d + \delta) = \begin{pmatrix} d_E & s_\lambda^{-1} \\ s_\lambda & -d_E \end{pmatrix} = \begin{pmatrix} I & 0 \\ -d_E s_\lambda & I \end{pmatrix} \begin{pmatrix} d_E & s_\lambda^{-1} \\ s_\lambda & 0 \end{pmatrix}$$

and so $\tau_\lambda = |\det(s_\lambda)_+ \det(s_\lambda)_-^{-1}|$, where $(s_\lambda)_+$ is the even-dimensional part of s_λ, $(s_\lambda)_-$ the odd-dimensional part.

Now $\hat{r}^*$ is a cochain map that induces $\beta: H^*(Y,Z;E) \to H^*(Y,Z;E)$. So one can compute $L(\beta^k) = L(\hat{r}^{*k})$ and so (D) gives $\varsigma_\beta = \det I - t(\hat{r}^*)_- / \det I - t(\hat{r}^*)_+$. Thus $|\varsigma_\beta(\lambda)| = \tau_\lambda^{-1}$ whenever s_λ is invertible.

Consider the long exact sequence of the triple $(\hat{M}, N \cup Y, N)$. One has isomorphisms

$$H^*(N \cup Y, N; \hat{E}) \simeq H^*(Y,Z;E) \simeq H^{*+1}(\hat{M}, N \cup Y; \hat{E})$$

for which the coboundary of the triple becomes $\pm(I-\beta)$. Thus $\hat{E}$ is acyclic for $(\hat{M}, N) \Leftrightarrow (I-\beta)$ is invertible.

When $I-\beta$ is invertible, we use the continuity of τ on the family $\hat{E} \otimes \pi^* E_\lambda$ (see §2) to see that $\tau(B, b^-; E) = \tau(\hat{M}, N; \hat{E}) = \lim_{\lambda \to 1} \tau_\lambda = |\varsigma_\beta(1)|^{-1}$. This proves Theorem 3.2 and concludes our discussion of steps (1), (2) and (3).

Finally suppose R is circular and E is a flat bundle over X. Then we take G as for step (3). There are finitely many points $x_j < x_{j+1}$ in $\mathbf{R} - G(R)$ such that $G^{-1}[x_j, x_{j+1}]$ isolates an invariant set I_j so that $I = \cup I_j$ is close as we please to R. In particular we can make I circular. Then if B_j is an isolating block for I_j then we have

$$\tau(B_j, b_j^-; E|B_j) = |\varsigma_j(t)|^{-1}\Big|_{t=1}$$

where ς_j is the twisted Lefschetz function for (B_j, b_j^-) and E, $\varsigma_j = \varsigma_{\beta_j}$, provided $\beta_j - id$ is invertible.

Smoothing the level set $G^{-1}x_j$ by sliding it along the flow, one obtains a smooth region X_j with

$$G^{-1}(-\infty, x_j - \epsilon) \subset X_j \subset G^{-1}(-\infty, x_j + \epsilon)$$

such that ϕ is transverse to ∂X_j. One has that $X_{j+1} - X_j$ isolates I_j (if ϵ is small). Then

$$\begin{aligned}\tau(X, \partial_- X; E) &= \prod_j \tau(X_{j+1}, X_j; E)\\ &= \prod_j \tau(B_j, b_j^-; E|B_j)\end{aligned}$$

Thus one has

Theorem 3.3 *$\tau(X, \partial_- X; E) = |\varsigma(t)|^{-1}\Big|_{t=1}$ where $\varsigma(t) = \prod_j \varsigma_j(t)$ and E is acyclic for all the pairs (X_{j+1}, X_j).*

This expression for R-torsion as the special value of a zeta function defined by the periodic orbits of a flow will be generalized in the next section. We mention a particular case of it that generalizes Theorem 3.1. Suppose that ϕ is a nonsingular Smale flow so that R is hyperbolic and 1-dimensional. Then R is circular and isolated. In this case $\varsigma(t)$ can be computed using a fine Markov family of local sections, signed transition matrices, and a corresponding cocycle for E. As shown in [F1] one recovers Franks' theorem relating the Alexander polynomial of a link to dynamics. When R is finite, so ϕ is NAMS, the result takes the simple form of Theorem 3.1.

When ϕ is an Axiom A-No Cycles flow that is homology finite (i.e. each homology class contains only finitely many closed orbits) then there is an improvement of Theorem 3.3 valid for any acyclic E. In [F1], Theorem 10 the case $\dim E = 1$ is developed (using the ALEX formalism).

§4. Fuller measures

Our goal now is to extend the purely topological results of earlier sections to certain other flows. As motivation, consider a closed Riemannian manifold X and its unit sphere bundle $M = SX$. By lifting the unit-speed geodesics in X one obtains a flow on M called *geodesic flow* whose periodic orbits correspond to closed geodesics in X. For geodesic flow, $R = M$ is not circular (for $\dim X > 1$) and so the methods of §3 don't apply. But in later sections we will show that for many X a properly formulated extension of the results of §3 does hold, so closed geodesics can be counted by R-torsion.

Let M be a smooth compact manifold and ϕ a smooth local flow on M transverse to the boundary of M. We will study the *periodic set* $\mathcal{P} = \mathcal{P}^\phi$ in space-time $M \times (0, \infty)$ consisting of pairs (x, t) with $\phi_t x = x$. For such a point we call $\gamma = \{(\phi_s x, t) | 0 \leq s \leq t\} \subset \mathcal{P}$ the periodic orbit (or closed orbit) determined by (x, t), so $\mathcal{P}$ is partitioned into closed orbits of ϕ. The time t will be called the period or length of γ and denoted $\ell(\gamma)$. The collection of these lengths $\{\ell(\gamma)\} = L_\phi = L$ will be called the *length spectrum* of ϕ, following the terminology used for geodesic flows. For any smooth nonsingular ϕ on a closed manifold, L is bounded away from 0.

In most cases of interest L is discrete. This holds for instance if ϕ is geodesic flow for a Riemannian manifold with an analytic metric [So]. By the variational description of closed geodesics as extrema of an energy functional, the energy will not vary over a smooth path of closed geodesics, hence neither will the length. The analyticity of the metric implies that $\mathcal{P}$ is an analytic subset of $M \times (0, \infty)$, hence that nearby points of $\mathcal{P}$ are connected by smooth paths. Thus L is indeed discrete. We call a closed orbit γ *isolated* if it is open in $\mathcal{P}$ in the induced topology on $\mathcal{P}$, i.e. if (x', t') is in $\mathcal{P}$ and near $(x, t) \in \gamma$ then $t' - t$ and $x' = \phi_s x$. If the underlying trajectory in M is an isolated invariant set for ϕ then γ is an isolated closed orbit but be warned that the converse fails (consider, say, the suspension flow of the flip map $\bar{z}$ on S^1 with γ the multiplicity one closed orbit through a fixed point). Of course L is also discrete if all closed orbits are isolated, e.g. for geodesic flow on a negatively curved manifold or for a generic smooth nonsingular flow (by the Kupka-Smale Theorem). When L is discrete the level sets $\mathcal{P}_\ell = \mathcal{P} \cap M \times \{\ell\}$ are open in $\mathcal{P}$ for all ℓ.

To a point $(x, t) \in \mathcal{P}$ we associate the closed trajectory $\phi_s x$, $0 \leq s \leq t$, and denote the corresponding free homotopy class by $c = c(x, t)$. When M is connected one can identify c with a conjugacy class in $\pi_1(M, b)$ by joining x to a basepoint b by some path. Note that c is locally constant on $\mathcal{P}$. Given a free homotopy class c and $\ell > 0$, we let $\mathcal{P}_{\ell,c}$ be the set of $(x, \ell) \in \mathcal{P}$ with $c(x, \ell) = c$. Then $\mathcal{P}_{\ell,c}$ is open in $\mathcal{P}$ provided L is discrete. Also $\mathcal{P}_{\ell,c}$ is compact.

We recall how to any compact open set $I \subset \mathcal{P}$ one can associate a *Fuller index* $\text{ind}_F(\phi, I) = \text{ind}_F(\phi) \in \mathbf{Q}$.

To begin, suppose $I = \gamma$ is an isolated closed orbit. For $(x, t) \in \gamma$ one chooses a codimension one submanifold S through x transverse to the flow ϕ, called a (local) Poincaré section. By following the trajectory of a point $s \in S$ near x for a time $t(s)$ near t, one again returns to S at a point $r(s)$. This local *return map* r sends a neighborhood S' of x in S to S and fixes x. As γ is isolated, there are no other fixed points near x. Thus there is a Lefschetz index $\text{ind}_L(r, x) \in \mathbf{Z}$ that measures the topological intersection of the graph of r and the diagonal in $S \times S$ at the point (x, x). As this index is independent of the choice of S and x, we denote it $\text{ind}_L(\gamma)$. Another integer associated to γ is its *multiplicity* $m = m(\gamma) > 0$ defined by the rule that t/m is the least period of x for ϕ, where $(x, t) \in \gamma$. The ratio $\text{ind}_L(\gamma)/m(\gamma)$ is the Fuller index $\text{ind}_F(\gamma)$.

For general I, one chooses an open set $\mathcal{O} \subset M \times (0, \infty)$ with $\mathcal{O} \cap \mathcal{P} = I$ and an open C^1 neighborhood U of ϕ so that for $\psi \in U$, $\mathcal{P}^\psi$ is disjoint from the boundary of $\mathcal{O}$. By choosing

$\psi \in U$ generically, the Kupka-Smale theorem assures $\mathcal{P}^{\psi} \cap \mathcal{O}$ will be a union of finitely many isolated closed orbits of ψ. The sum of their Fuller indices is defined to be $\mathrm{ind}_F(\phi, I)$. Of course it must be checked that this definition is independent of the choices made [Fu2,MP]. Indeed, the Fuller index enjoys the property of homotopy invariance: given an open set $\mathcal{O} \subset M \times (0, \infty) \times [0,1]$ and a continuous family ϕ^s of flows, $0 \leq s \leq 1$, such that for no $(x,t,s) \in \partial\mathcal{O}$ is $(x,t) \in \mathcal{P}^{\phi^s}$, the Fuller index of $\{(x,t) \in \mathcal{P}^{\phi^s} | (x,t,s) \in \mathcal{O}\}$ is constant in s. Moreover ind_F is additive: if I, J are compact-open subsets of $\mathcal{P}^{\phi}$ then

$$\mathrm{ind}_F(I \cup J) = \mathrm{ind}_F(I) + \mathrm{ind}_F(J) - \mathrm{ind}_F(I \cap J)$$

Given a flow ϕ on M with discrete length spectrum we define the *Fuller measure* μ_ϕ to be the atomic measure on $(0, \infty)$ with support $(\mu_\phi) \subset L$ so that the mass of μ_ϕ at $\ell \in L$ is $\mathrm{ind}_F(\phi, \mathcal{P}_\ell)$, i.e.

$$\mu_\phi = \sum_{\ell \in L} \mathrm{ind}_F(\phi, \mathcal{P}_\ell)\delta_\ell$$

Given a flat bundle E over M and a free homotopy class c, let $\chi(c)$ be the trace of the holonomy of E about a representative of c. The *twisted Fuller measure* of ϕ relative to E is defined to be

$$\mu_{\phi,E} = \sum_{\ell \in L} \sum_{c} \mathrm{ind}_F(\phi, \mathcal{P}_{\ell,c})\chi(c)\delta_\ell$$

Note that the masses are topological but their positions depend on the parametrization of ϕ.

Suppose I is an isolated invariant set for ϕ and $\mathcal{O}$ is an isolating neighborhood for I. Then the intersection $\mathcal{P} \cap I \times (0, \infty)$ is open in $\mathcal{P}$ and we set

$$\mu_{\phi|\mathcal{O},E} = \sum_{\ell \in L} \sum_{c} \mathrm{ind}_F(\mathcal{P}_{\ell,c} \cap I \times (0,\infty))\chi(c)\delta_\ell$$

i.e. we restrict to closed orbits whose trajectories are contained in $\mathcal{O}$. If in a certain context only one bundle E or one flow ϕ is under consideration, we may omit that subscript.

The twisted Fuller measure has a number of formal properties that correspond to the properties of R-torsion discussed in §1. Suppose first that $A, B \subset X$ are compact regions with smooth boundary transverse to ϕ and that ϕ flows into A, B on $\partial A, \partial B$. Then if ∂A is transverse to ∂B one has for any E

$$\mu_{\phi|A\cup B} = \mu_{\phi|A} + \mu_{\phi|B} - \mu_{\phi|A\cap B} \qquad (I)$$

To see this, note that the set $\mathcal{P} \cap ((A \cup B) \times (0, \infty))$ is partitioned into the *open* subsets $\mathcal{P} \cap ((A - B) \times (0, \infty))$, $\mathcal{P} \cap ((B - A) \times (0, \infty))$ and $\mathcal{P} \cap ((A \cap B) \times (0, \infty))$ since any closed orbit that meets A (or $B, A \cap B, A \cup B$) lies in its interior for all time. Intersecting these sets with $\mathcal{P}_{c,\ell}$ and using the additivity of ind_F proves (I). Note, however, that (I) would not hold if A, B were arbitrary isolating neighborhoods since then one might have a closed orbit in $A \cup B$ that was not entirely contained in either A or B.

Next, suppose A, B are M-blocks and that A is a squeeze or shave of B. Then trivially

$$(II) \qquad \mu_{\phi|A} = \mu_{\phi|B}$$

Next consider the case $X = \mathbf{R}/\mathbf{Z}$, ϕ the natural $\mathbf{R}$-action on X. Then if $t(x) = x + 1$ is the preferred generator of $\pi_1 X$ and $A = \rho_E(t)$ one clearly has

$$(III) \qquad \mu_\phi = \sum_{n>0} \frac{1}{n} \operatorname{Tr}(A^n)\delta_n$$

Now suppose $\pi: M \to B$ is a smooth bundle with fiber F, with M, B, F smooth, closed manifolds. Suppose ψ is a smooth flow on M, ϕ a smooth flow on B and $\pi\psi_t = \phi_t\pi$ for all t, i.e. ϕ is a smooth factor of ψ. If ϕ has discrete length spectrum so does ψ and for any flat bundle E over B

$$(IV) \qquad \mu_{\psi,\pi^* E} = \sum (-1)^i \mu_{\phi, E\otimes H^i F}$$

To see this one first considers the case when each closed orbit γ of ϕ is isolated. Then one can partition $\mathcal{P}^\psi$ into open sets by assigning to each closed orbit of ψ the underlying closed orbit of ϕ, i.e. applying $\pi \times id: \ E \times (0, \infty) \to B \times (0, \infty)$ to $\mathcal{P}^\psi$ and taking the inverse images $\mathcal{O}_\gamma \subset \mathcal{P}^\psi$ of the components γ of $\mathcal{P}^\phi$. We will show that the contribution of γ to the right hand side equals the contribution of $\mathcal{O}_\gamma$ to the left. Taking the coefficient of $\delta_{\ell(\gamma)}$, this reduces to

Lemma 4.1 $\mathrm{ind}_F(\psi, \mathcal{O}_\gamma) = L(r)\ \mathrm{ind}_F(\phi, \gamma)$ *where r is the monodromy of π over the trajectory in X determined by γ.*

To prove this, let $\delta(s) = \phi_s(x)$, $0 \le s \le t/m(\gamma)$, where $(x, t) \in \gamma$. Choose B_0 a tubular neighborhood of δ, i.e. a disc bundle over S^1 with the dimension of X equipped with an immersion into X so that a section $\gamma_0 \subset B_0$ traces δ. Pulling back gives an F-bundle $\pi_0: M_0 \to B_0$ and vector fields Z_0, Y_0 on M_0, B_0 that correspond to the generators Z, Y of

ψ, ϕ. In this way the lemma reduces to the case when γ covers an embedded circle $\delta \subset B$ with multiplicity $m(\gamma)$.

Assuming this holds, we perturb ψ to $\hat{\psi}$ so that $\pi\hat{\psi}_t = \phi_t\pi$ and the orbits of $\hat{\psi}$ over γ are isolated, so that $\text{ind}_F(\psi, \mathcal{O}_\gamma)$ is a sum over such orbits $\hat{\gamma}$ of terms $\text{ind}_F(\hat{\psi}, \hat{\gamma})$. We choose the monodromy r to be $\psi_{\ell(\gamma)}|F_x$, F_x the fiber over a point x in the trajectory of γ. Using the Lefschetz fixed point formula, $L(r)$ is also a sum over $\hat{\gamma}$ of terms corresponding to the intersections of F with the trajectory of $\hat{\gamma}$. We will show that a given $\hat{\gamma}$ contributes equally to both sides, proving the Lemma.

The multiplicity $m(\hat{\gamma})$ divides $m(\gamma)$ and $\hat{\gamma}$ determines $m(\gamma)/m(\hat{\gamma})$ fixed points of r, each of the same Lefschetz index λ. Take a Poincaré section $S_B \subset B$ for γ and lift it to a Poincaré section $S_E \subset E$ for $\hat{\gamma}$. Then the local return map r_B for S_B is a factor of the local return map r_E for s_E with r the restriction of r_E to F_x. We use the following easy result (whose proof we omit) with $f = r_E, f_1 = r_B, f_2(0, y) = r$:

Lemma 4.2 *If $f(x, y) = (f_1(x), f_2(x, y))$ is a fiber preserving map on a neighborhood of $(0,0) \in \mathbf{R}^m \times \mathbf{R}^n$ with $(0,0)$ as an isolated fixed point then $0 \in \mathbf{R}^m$ is an isolated fixed point of f_1, $0 \in \mathbf{R}^n$ is an isolated fixed point of $f_2(0, y)$ and*

$$\text{ind}_L(f(x, y), (0,0)) = \text{ind}_L(f_1(x), 0)\text{ind}_L(f_2(0, y), 0)$$

In our setting this gives $\text{ind}_L(\hat{\gamma}) = \text{ind}_L(\gamma) \cdot \lambda$. Summing this over the $m(\gamma)/m(\hat{\gamma})$ fixed points and dividing by $m(\gamma)$ gives

$$\text{ind}_F(\hat{\psi}, \hat{\gamma}) = \text{ind}_F(\phi, \gamma) \cdot \lambda \cdot m(\gamma)/m(\hat{\gamma})$$

As $\sum_{\hat{\gamma}} \lambda m(\gamma)/m(\hat{\gamma}) = L(r)$ by the Lefschetz fixed point formula, Lemma 4.2 follows and so does formula (IV).

The case when ϕ is trivial, B is connected and ψ has discrete length spectrum gives, for E a bundle on M

$$(V) \qquad \mu_\psi = \chi(B)\mu_{\psi|F}$$

Note that ψ is tangent to fibers. A proof can be given by fixing an $\ell \in L^\psi$, choosing a flow $\hat{\psi}$ near ψ with isolated closed orbits that induces a gradient flow $\hat{\phi}$ in the base and again using the preceding Lemma 4.2 to measure the contribution of $\mathcal{P}_\ell$ to both sides.

When F is finite, L_ϕ is discrete iff L_ψ is discrete. In this case one can perturb ϕ to a flow with isolated closed orbits, lift the perturbation to perturb ψ and compute

$$(VI) \qquad \mu_\psi(E) = \mu_\phi(\pi_* E)$$

Thus all 6 properties of R-torsion are mirrored in properties of (twisted) Fuller measure. It will be shown later that the "total mass" of $\mu_\phi(E)$, suitably interpreted, is sometimes $-\log \tau_E$. This is motivated by the strong analogy in their formal properties. The homotopy invariance of the Fuller index implies that if one takes a family of flows (each with discrete L) the masses of the corresponding Fuller measures will move like beads on a wire: no mass can be lost except by beads sliding off the end, i.e. by orbits whose periods tend to infinity. Aside from such catastrophic losses (which do occur, as examples will show) the "total mass" of $\mu_\phi(E)$ is *formally* invariant under deformation, and so should be a topological quantity.

One property of $\mu_\phi(E)$ that deserves special attention is its connection to the distributional traces of certain natural linear operators. Suppose ϕ is a smooth nonsingular flow on a closed manifold X. Each closed orbit γ gives a curve of intersection of the graph of ϕ $\{(x, t, \phi_t x)\}$ and the graph of the trivial flow $\{(x, t, x)\}$ in $X \times (0, \infty) \times X$. We call γ *simple* if this intersection is transverse along γ. In terms of the return map r_γ to a Poincaré section for $(x, t) \in \gamma$, γ is simple if the differential $T_x r_\gamma$ has no nonzero fixed vector. If all closed orbits γ of ϕ are simple then L is discrete. Moreover the family $\phi_t^*, t > 0$, of linear operators induced by ϕ_t acting on functions has a trace in a suitable sense [GS]. This trace is a measure ν_ϕ on the parameter space $t > 0$ supported on L. Each γ contributes a term $\ell_0 \frac{1}{|\det I - T_x r_\gamma|} \delta_{\ell(\gamma)}$ to ν_ϕ where $\ell_0(\gamma) = \ell(\gamma)/\mu(\gamma)$. In case ξ is a vector bundle over X equipped with a lift of the flow ϕ_t to linear bundle maps ψ_t, one has a similar trace ν_ψ for the action ψ_t^* on the sections of ξ, where each γ now contributes

$$\ell_0(\gamma) \frac{\text{Trace } h_\psi(\gamma)}{|\det I - T_x r_\gamma|} \delta_{\ell(\gamma)}$$

to ν_ψ, where $h_\psi(\gamma)$ is the holonomy of ξ around the trajectory of γ. When $\xi = E \otimes \Lambda^i TM$ with $0 \leq i \leq \dim M$ and E a flat bundle, there is a natural choice for ψ given by parallel transport in the flat connection tensored with $\Lambda^i T\phi_t$: here we just write $\nu_\phi(E, i)$ for ν_ψ and $h_{E,i}(\gamma)$ for $h_\psi(\gamma)$.

The relation of $\mu_\phi(E)$ to the traces $\nu_\phi(E,i)$ is the following identity, for ϕ with simple closed orbits:

$$(VII) \qquad t\mu_\phi(E) = \sum(-1)^{i+1}\, i\nu_\phi(E,i)$$

To see this, we check that each closed orbit γ contributes the same on both sides, i.e.

$$\ell(\gamma)\cdot\chi(\gamma)\frac{\mathrm{ind}_L(r_\gamma,x)}{m(\gamma)} = \sum(-1)^{i+1}\, i\ell_0(\gamma)\frac{\chi(\gamma)\mathrm{Tr}\Lambda^i T_x\phi_{\ell(\gamma)}}{|\det I - T_x r_\gamma|}$$

This follows from the formulas

(a) $\mathrm{ind}_L(r_\gamma,x) = \mathrm{sgn}\ \det I - T_x r_\gamma$

(b) $\det I - T_x r_\gamma = \sum(-1)^i\ \mathrm{Tr}\ \Lambda^i T_x r_\gamma$

(c) $\mathrm{Tr}\ \Lambda^i T_x\phi_{\ell(\gamma)} = \mathrm{Tr}\ \Lambda^i T_x r_\gamma + \mathrm{Tr}\Lambda^{i-1}T_x r_\gamma$

in which (a) is the usual formula for Lefschetz index at a simple fixed point, (b) is a linear algebra identity and (c) follows from the fact that $T_x\phi_{\ell(\gamma)}$ has a fixed vector tangent to ϕ and acts like $T_x r_\gamma$ on the quotient. Multiply (c) by i and take an alternating sum to get

$$\sum(-1)^{i+1}\ i\ \mathrm{Tr}\ \Lambda^i T_x\phi_{\ell(\gamma)} = \sum(-1)^i\ \mathrm{Tr}\ \Lambda^i T_x r_\gamma$$

and then use (b) and (a).

Formula (VII) provides a link between Fuller measures and linear operators. It helps to motivate the definition of $\mu_\phi(E)$: it is that combination of the traces $\nu_\phi(E,i)$ in which all the masses are rational multiples of the $\chi(\gamma)$. It suggests much more, namely a connection with analytic torsion (see §6) but this connection is now understood only in the special case of geodesic flows on hyperbolic manifolds [F3] and flat manifolds (§6).

§5. R-torsion as a special value

We will present a zeta function that often enables one to express R-torsion in terms of the twisted Fuller measure. Let M be a manifold with a flat bundle E and ϕ a smooth flow on M with discrete length spectrum. We define the *torsion function* as a formal Dirichlet series

$$Z_{\phi,E}(s) = \exp -\mathfrak{L}\mu_{\phi,E}(s)$$

where for a measure μ on $(0,\infty)$, $\mathfrak{L}\mu$ denotes the Laplace transform

$$\mathfrak{L}\mu(s) = \int_0^\infty e^{-ts} d\mu(t)$$

Typically $\mu_{\phi,E}$ will have some exponential bound so that $Z_{\phi,E}(s)$ is a holomorphic function of s for Re s large. Assume that this is so and that $Z_{\phi,E}$ admits a meromorphic continuation to some region containing $[0,\infty)$. Then we let $\Psi_{\phi,E} = |Z_{\phi,E}(0)| \in [0,\infty]$. Note that formally

$$-\log \Psi_{\phi,E} = \text{Re} \int_0^\infty d_{\mu_{\phi,E}}(t)$$

so $\Psi_{\phi,E}$ measures the total twisted Fuller index.

Now suppose E has a flat density and is acyclic. If the special value $\Psi_{\phi,E}$ equals τ_E we say that the pair (ϕ, E) is *Lefschetz*. In all known cases when τ_E can be computed from the closed orbits of a flow, one can find a time parametrization ϕ so that (ϕ, E) is Lefschetz. In some cases this is the only known way to express τ_E in terms of closed orbits. At the end of this section some pairs (ϕ, E) will be given that are not Lefschetz so the problem is to find natural Lefschetz pairs.

To get an idea of how $\Psi_{\phi,E}$ can be computed, consider the simple case $M = S^1$, ϕ a nonsingular flow of least period p. Then for $A = \rho_E(t)$ as before,

$$\begin{aligned} Z_{\phi,E}(s) &= \exp - \sum_{m>0} \frac{1}{m}\text{Tr}(A^m)e^{-mps} \\ &= \det I - Ae^{-ps} \end{aligned}$$

for Re s large enough so $|e^{ps}|$ exceeds the spectral radius of A. Clearly $Z_{\phi,E}$ is entire in s and $\Psi_{\phi,E} = |\det I - A|$. By (III) of §1 this is indeed τ_E.

It follows formally from formula (V) of §1 and 4 that for the periodic flow on a circle bundle any flat bundle E with $\det I - \rho_E(t) \neq 0$ is Lefschetz. We will extend this result to a locally free circle action and even to a flow that commutes with such an action. A nice example to keep in mind is the flow on a lens space induced by the scalar action of S^1 on its universal cover: this leads to a dynamic interpretation of the formulas of Reidemeister and Franz.

Let ϕ_t be a smooth locally free S^1 action on M and ψ_s a smooth nonsingular flow on M that commutes with ϕ_t. One has an orbifold quotient M/S^1 which is locally modeled on the quotient of Euclidean space by a finite cyclic group, and ψ_s induces a smooth flow $\overline{\psi}_s$ on this quotient. One can mimic the proof of the Kupka-Smale theorem on this orbifold to perturb $\overline{\psi}$ to a flow with only hyperbolic stationary points and closed orbits. This perturbation $\overline{\psi}^*$ of $\overline{\psi}$ can be lifted to a perturbation ψ^* of ψ. A degenerate family of closed orbits of ψ^* that covers a closed orbit of $\overline{\psi}^*$ has Fuller index 0 and so contributes nothing to the Fuller measures of ψ^*. If L_ψ is discrete then ψ^* can be chosen with $L_{\psi^*} \subset L_\psi$. The Fuller measures of ψ^* will only contain contributions from the hyperbolic closed orbits γ that cover stationary points of $\overline{\psi}^*$, and will equal the corresponding measures for ψ.

Fix γ and a Poincaré section S at a point $x \in \gamma$. Then we have return maps r_{ψ^*} and r_ϕ defined near x in S and these commute. Let X be the vectorfield obtained by projecting $\frac{d\psi^*}{dt}$ into S along $\frac{d\phi}{dt}$. If the generic orbit of ϕ_t has periods 1 (as we may assume) and the ϕ_t orbit through x has period $\frac{1}{m}$ then one can choose local coordinates on S so that $x = 0$ and r_ϕ is a linear map of order m. The differential $T_0 r_{\psi^*}$ commutes with r_ϕ and is hyperbolic. Thus we can expand $T_0 r_{\psi^*} = A_u \oplus A_s$, $r_\phi = R_u \oplus R_s$ with A_u an expansion and A_s a contraction. If $B = T_0 X$ then $B = B_u \oplus B_s$ and $e^B = (T_0 r_{\psi^*}) r_\phi^{\pm 1}$ where the exponent depends on whether the flows ψ^* and ϕ are parallel or opposite on γ. It follows that $\Delta(\gamma) = \operatorname{sgn} \det A_u = \operatorname{sgn} \det R_u^{\mp 1} e^{B_u} = \operatorname{sgn} \det R_u$.

Suppose that $\rho_E(t)$ does not have 1 as an eigenvalue, where t denotes the homotopy class of the generic orbit of ϕ. Suppose, seeking a contradiction, that $\Delta(\gamma)$ is an eigenvalue of $\rho_E(\gamma)$. Since $\rho_E(\gamma)^m = \rho_E(t)$, one must have $\Delta(\gamma) = -1$. Thus -1 is an eigenvalue of R_u. Choose a (-1)-eigenvector for R_u and let δ be the corresponding orbit of ϕ. Then $\rho_E(\delta) = \rho_E(\gamma)^2$ has 1 as an eigenvalue, which is a contradiction. Thus we have

$$\det I - \Delta(\gamma)\rho_E(\gamma) \neq 0$$

Now consider the contribution to $\mu_{\psi^*,E}$ of the multiples of γ. The Lefschetz index of γ^n is

$(-1)^{u(\gamma)}\Delta(\gamma)^n$ and so

$$\mu_{\psi,E} = \mu_{\psi^*,E} = \sum_{\gamma}\sum_{n>0}((-1)^u\Delta(\gamma)^n/n)\delta_{n\ell(\gamma)}(\mathrm{Tr}\,\rho(\gamma)^n)$$

$$Z_{\psi,E}(s) = \prod_{\gamma}(\det I - \Delta(\gamma)\rho(\gamma)e^{-s\ell(\gamma)})^{(-1)^{u(\gamma)}}$$

$$\Psi_{\psi,E} = \prod_{\gamma}|\det I - \Delta(\gamma)\rho(\gamma)|^{(-1)^{u(\gamma)}}$$

Now we take $\psi = \phi$ and $\overline{\psi}^*$ a time 1 map of a gradient flow on M/S^1 so that ψ^* is a NAMS flow. Theorem 3.1 then gives E acyclic and $\Psi_{\phi,E} = \tau_E$, i.e. (ϕ, E) is Lefschetz.

For arbitrary ψ and some choice of ψ^* as above, use $\phi_s^* = \phi_s\psi_{\epsilon s}^*$, ϵ a small positive constant. Then the above formula for $\Psi_{\psi,E}$ shows $\Psi_{\psi,E} = \Psi_{\psi^*,E} = \Psi_{\phi^*,E} = \Psi_{\phi,E}$. Thus (ψ, E) is also Lefschetz. We summarize this in

Theorem 5.1 *Suppose one has a smooth, locally free S^1 action on a closed manifold M and E is a flat bundle over X with a flat density such that $1 \notin \mathrm{Spec}\,\rho_E(t)$, t the class of a generic S^1-orbit. Then E is acyclic. If ψ is a smooth flow commuting with this circle action and ψ has discrete L then (ψ, E) is Lefschetz.*

In the course of the proof, we essentially proved

Theorem 5.2 *Suppose ϕ_t is a NAMS flow on M and that for each closed orbit $\gamma, \Delta(\gamma) \notin$ Spec $\rho_E(\gamma)$. Then (ϕ, E) is Lefschetz.*

Of course the latter is just a rephrasing of Theorem 3.1. We can similarly rephrase the other results of §3.

We will mention a practical way of computing Fuller indices in many cases. Recall that a submanifold Z of fixed points of a diffeomorphism f is *clean* if for each $z \in Z$ the subset of T_zM fixed by T_zf lies in (hence equals) T_zZ. If moreover for all z T_zZ is the largest subset on which T_zf is unipotent we will say Z is *immaculate*. (Note that if Z is a point clean and immaculate just mean the fixed point is simple). Then at each $z \in Z$ the differential induces a normal derivative N_zf on (T_zM/T_zZ) that fixes no nonzero vector. If Z is connected then sgn $(\det I - N_zf)$ is constant and will be denoted $\sigma = \sigma(Z) \in \{\pm 1\}$.

Suppose $Z \subset \mathrm{Fix}\,\phi_t$, $t > 0$, is immaculate and connected. One can compute the Fuller index of the subset $Z \times \{t\} \subset \mathcal{P}$ as follows. Note that Z carries a natural locally free action of $\mathbf{R}/t\mathbf{Z} \simeq S^1$ induced by ϕ. The orbit space Z/S^1 is a smooth orbifold and as such has an Euler characteristic $\chi \in \mathbf{Q}$. Let $m \in \mathbf{Z}^+$ be the multiplicity of a generic orbit for $\phi_t|Z$. Then

Lemma 5.3 $\mathrm{ind}_F(\phi, Z \times \{t\}) = \sigma\chi/m$.

For a proof one simply perturbs the S^1 action on Z to a Morse-Smale flow ψ^* (as above, with $\psi = \phi$) and extends to a perturbation of ϕ on a neighborhood of Z that we will denote by $\hat{\psi}$. Then one can compute the contribution of a closed orbit γ of $\hat{\psi}$ to χ and to the Fuller index. If the isotropy subgroup of $\mathbf{R}/t\mathbf{Z}$ has order n at a point $z \in \gamma$ then m divides n and γ contributes $\frac{m}{n}(-1)^u$ to χ, where u is the unstable dimension of γ as a closed orbit of $\psi^* = \hat{\psi}|Z$, and contributes $\frac{1}{n}(-1)^u\sigma$ to the Fuller index. Summing over γ proves the Lemma.

For an example of a non-Lefschetz pair (ϕ, E) take an automorphism A of T^2 with eigenvalues $\alpha, \alpha^{-1}, \alpha > 1$, and form the mapping torus M_A. Let ϕ be the horocycle flow along the eigendirection for α. Then ϕ has no closed orbits so $\Psi_{\phi,E} = 1$ for all E. But if $E = E_\lambda$, $|\lambda| = 1, \lambda \neq 1$, then by §1, E is acyclic and $\tau(E) = |1-\lambda|^2/|1-\lambda\alpha||1-\lambda\alpha^{-1}| = \frac{2-(\lambda+\lambda^{-1})}{(\alpha+\alpha^{-1})-(\lambda+\lambda^{-1})} \neq 1$. Note, however, that ϕ can be approximated by flows with a constant nonzero component transverse to the fibration over S^1 and these nearby flows are Lefschetz. Using the plugged flows of Wilson and Schweitzer [W, Sch] one can produce non-Lefschetz pairs (ϕ, E) for any acyclic E with $\tau(E) \neq 1$, where ϕ has no stationary points and no closed orbits (note E acyclic $\Rightarrow \chi(M) = 0 \Rightarrow M$ has a nonsingular flow) provided dim $M \geq 3$. It is clear, then, that flows without closed orbits are pathological in this respect, and some restrictions on ϕ must be made.

In the next two sections we restrict our attention to geodesic flows. If the geodesic flow ϕ_t on $SX = M$ and the bundle E on M form a Lefschetz pair we will say that (M, E) is Lefschetz. If $E = \pi^*E'$, where E' is a bundle on X and $\pi: M \to X$ is the obvious projection, we say also that (X, E') is Lefschetz. If in a certain context M is fixed we say just that E or E' is Lefschetz.

§6. Manifolds of constant curvature

We will show that various flat bundles on constant curvature manifolds are Lefschetz. In the case of positive curvature topological methods suffice but for curvature zero or negative analytic tools are needed, particularly the Selberg trace formula (STF) and the analytic torsion theorem of Cheeger and Müller.

If X has constant positive curvature, we can rescale the metric without changing Ψ so as to assume the curvature is $+1$. Then $X = \Gamma\backslash S^n$ where $\Gamma \subset 0(n+1)$. If n is even then $X = \mathbf{R}P^n$ or S^n and no flat bundle E is acyclic. Thus we suppose n odd and $\Gamma \subset SO(n+1)$.

More generally, suppose that $X = \Gamma\backslash\tilde{X}$ where $\tilde{X}$ is a finite regular cover and Γ is a fixed point free subgroup of the connected component $I_0(\tilde{X})$ of the compact Lie group $I(\tilde{X})$ of isometries of $\tilde{X}$. Choose a representation $\rho: \Gamma \to Gl(m, \mathbf{C})$ and let $E(\rho)$ be the corresponding flat bundle on X. We will show

Theorem 6.1 *If no nonzero vector in* $\mathbf{C}^m$ *is fixed by* ρ *then* $E(\rho)$ *is acyclic and Lefschetz.*

Taking $\tilde{X} = S^n$, n odd, gives

Corollary 6.2 *If* X *has constant positive curvature then any acyclic bundle over* X *is Lefschetz.*

Note that since $\Gamma \subset I_0(\tilde{X})$, Γ fixes the cohomology of $\tilde{X}$. Thus every nontrivial irreducible representation of Γ is acyclic and the acyclicity result of Theorem 6.1 follows.

Now we prove the theorem for cyclic Γ. Say g generates Γ. Then there is a compact one-parameter group $H \subset I_0(\tilde{X})$ containing g (recall that g belongs to a maximal torus and g has finite order). Since Γ acts without fixed points, the action of H on $\tilde{X}$ by isometries is locally free. So H acts on $S\tilde{X}$ and H/Γ acts on SX. Clearly this is a locally free circle action ϕ_s commuting with the geodesic flow ψ_t on SX. The hypothesis of Theorem clearly holds since the closed orbits of ϕ_s generate Γ and ρ does not contain the trivial representation. So Theorem 6.1 holds for Γ cyclic.

From this one can deduce the general case by induction. We need a result on characters of finite groups. Suppose that Γ is finite and $R(\Gamma)$ is the rational vector space spanned by the characters of Γ with the usual inner product. We let 1_Γ denote the trivial character of Γ and r_Γ the character of the regular representation. We let $V_\Gamma \subset R(\Gamma)$ be the subspace spanned by induced characters σ^Γ, where σ runs over nontrivial characters of cyclic subgroups C of Γ. We show

Lemma 6.3 $V_\Gamma = (1_\Gamma)^\perp$

First note that $\sum_{g\in\Gamma} \sigma^\Gamma(g) = [\Gamma : C] \sum_{g\in C} \sigma(g) = 0$, so $V_\Gamma \subset (1_\Gamma)^\perp$.

Next a theorem of Artin ([La], p. 484) implies that $R(\Gamma)$ is spanned by the σ^Γ coming from cyclic $C \subset \Gamma$ where σ is allowed to be trivial. Thus it suffices to show

$$(1_C)^\Gamma \in V_\Gamma + \mathbf{Q}1_\Gamma$$

Now $r_C - 1_C \in V_C$ ([La],pp. 476-7). Thus $(1_C)^\Gamma \in (r_C)^\Gamma + (V_C)^\Gamma$. By transitivity of induced characters, $(V_C)^\Gamma \subset V_\Gamma$ and $(r_C)^\Gamma = [(1_e)^C]^\Gamma = (1_e)^\Gamma = r_\Gamma$, with e the one element group. Thus $(1_C)^\Gamma \in r_\Gamma + V_\Gamma = 1_\Gamma + V_\Gamma$, proving the lemma.

Now for the flat bundle $E(\sigma^\Gamma)$ over $\Gamma\backslash\tilde{X}$ corresponding to the representation $\sigma^\Gamma \in V_\Gamma$, one knows that both Ψ and τ are the same as for a flat bundle $E(\sigma)$ over $C\backslash\tilde{X}$ corresponding to σ, by (VI) of §1, §4. Since $E(\sigma)$ is Lefschetz so is $E(\sigma^\Gamma)$. But then the lemma shows that for any ρ with no trivial component one has

$$m\rho = \oplus n_i(\sigma_i^\Gamma)$$

for some $m \in \mathbf{Z}^+$, $n_i \in \mathbf{Z}, \sigma_i \in \hat{C}_i$. This gives

$$m\mu_\rho = \sum n_i \mu_{\sigma_i^\Gamma}$$
$$\Psi_\rho^m = \prod_i \Psi_{\sigma_i^\Gamma}^{n_i} = \prod_i (\tau_{\sigma_i^\Gamma})^{n_i} = \tau_\rho^m$$

so that $\Psi_\rho = \tau_\rho$, as desired.

We will now consider the case of a flat Riemannian manifold $X(K = 0)$. We will use the theorem of Cheeger [Ch] and Müller [Mu] to express the R-torsion τ_E of an acyclic unitary E over SX in terms of the eigenvalues of certain Laplacians. Then by an application of the Selberg trace formula (STF) we will prove that E is Lefschetz.

We recall the *analytic torsion* T_ρ associated to a unitary representation $\rho: \pi_1 X \to U(m)$ of the fundamental group of a closed Riemannian manifold X. Let Δ denote the Laplacian on forms with coefficients in E_ρ, so $\Delta = \bigoplus_{p=0}^{n} \Delta_p$ where $n = \dim X$ and Δ_p denotes the restriction of Δ to twisted p-forms. Let β_p denote the dimension of $H^p(X; E_\rho)$. Then Seeley has shown that Tr Δ_p^{-s} has a meromorphic continuation $\varsigma_p(s)$ from large Re s to $\mathbf{C}$ which is regular at $s = 0$. The identity $(\sum_\lambda \lambda^{-s})'(0) = -\log \prod_\lambda \lambda$ for finite sums suggests the definition

$$\text{Det } \Delta_p = \exp -\varsigma_p'(0)$$

Ray and Singer defined $T_\rho \in (0, \infty)$ by

$$T_\rho^2 = \frac{(\mathrm{Det}\Delta_1)\ (\mathrm{Det}\Delta_3)^3 \cdots}{(\mathrm{Det}\Delta_2)^2\ (\mathrm{Det}\Delta_4)^4 \cdots}$$

They conjectured that if $H^p(X; E_\rho)$ is endowed with the density D_p associated to the Hodge inner product on harmonic p-forms that

$$\tau(E_\rho, D_p) = T_\rho$$

and this was proven by Cheeger and Müller (independently and by two quite distinct methods). In particular if E_ρ is acyclic

$$\tau(E_\rho) = T_\rho$$

Thus R-torsion is a spectral invariant of the Laplacian on forms in E_ρ.

We introduce a function

$$J(t) = \Sigma(-1)^{j+1}\ j \text{ Trace } e^{-t\Delta_j}, \quad t > 0$$

that is a convenient linear combination of the traces of heat kernels on j-forms. One has $J(t) = \Sigma c_\lambda e^{-t\lambda}$ for certain $c_\lambda \in \mathbf{Z}, \lambda \in \mathrm{Spec}\ \Delta$.

Lemma 6.4 *For X flat* $\sum\limits_{\lambda>0} \frac{|c_\lambda|}{\lambda} < \infty$

Granted this we will show any acyclic E_ρ over X is Lefschetz. We introduce a constant $a = m \cdot vol\ X$ if $n = 1$, $a = 0$ if $n > 1$. Then one has

$$(STF) \qquad J(t) = (4\pi t)^{-\frac{1}{2}}\left[a + \int_0^\infty \ell e^{-\ell^2/4t} d\mu_E(\ell)\right]$$

with μ_E the Fuller measure for the geodesic flow ϕ on SX. We leave the case $n = 1$ as an exercise and prove this formula for $n > 1$. First one computes μ_E. Each conjugacy class in $\pi_1 X$ is represented by a family of closed geodesics. If this family consists of an isolated geodesic γ (consider X=Klein bottle, for example) then its local Poincaré map has Lefschetz index $+1$, so μ_E receives a contribution of $a_\gamma = \frac{\ell(\gamma)}{m(\gamma)} \mathrm{Tr}\ \rho(\gamma)$ at the point $\ell(\gamma) \in (0, \infty)$. If this family is degenerate it defines a flat submanifold F of SX whose orbit space F/S^1 is a flat orbifold of positive dimension, hence has zero Euler characteristic: this can be used to compute that the Fuller index of F is zero, much as in Lemma 5.3. So μ_E only consists of contributions from isolated γ. Then (STF) becomes

$$J(t) = \left(\sum a_\gamma e^{-\ell(\gamma)^2/4t}\right)/\sqrt{4\pi t}$$

which was proven by Sunada and Urakawa for orientable X [SU]. Their proof works just as well for nonorientable X: one uses the explicit form of the heat kernel on Euclidean space and Selberg's method of expressing traces as sums over conjugacy classes. A key step in their computation that forces the cancellation of the contributions of nonisolated γ is the elementary identity (used in a different geometric setting in formula (VII),§4)

$$\sum(-1)^{p+1}\ p \operatorname{Tr} \Lambda^p A = \det I - B$$

valid for any $(n-1)$ by $(n-1)$ matrix B, with A an n by n matrix $\begin{pmatrix} B & 0 \\ * & 1 \end{pmatrix}$. This is applied to the holonomy $A(\gamma) = T_p\phi_{\ell(\gamma)}$, $p \in \gamma$ around the geodesic γ: note γ is isolated $\Leftrightarrow \det I - B \neq 0$.

Now we take the Laplace transform of (STF) at x^2 with $x >> 0$ and use an integration formula cited below to find

$$\sum_\lambda c_\lambda \frac{2x}{x^2+\lambda^2} = \int_0^\infty e^{-x^2 t} 2x J(t) dt = a + \frac{Z'(x)}{Z(x)}$$

This shows that $Z(x)$ has a meromorphic continuation to $\mathbf{C}$ since its logarithmic derivative has a meromorphic continuation to $\mathbf{C}$ (Lemma 6.4 implies that the series on the left converges uniformly for $x \neq \pm\lambda i$) with only simple poles and integer residues ($\frac{2x}{x^2+\lambda^2} = \frac{1}{x+\lambda i} + \frac{1}{x-\lambda i}$).

Next we introduce

$$\begin{aligned} F(t) &= (J(t) - c_0) + (c_0 - a(4\pi t)^{-\frac{1}{2}})\chi_{(0,1]} \\ &= (4\pi t)^{-\frac{1}{2}}\left[a\chi_{(1,\infty)} + \int_0^\infty \ell e^{-\ell^2/4t} d\mu_E(\ell)\right] - c_0\chi_{(1,\infty)} \end{aligned}$$

Note that the first formula for F shows it is exponentially small as $t \to \infty$ and the second formula shows that it is exponentially small as $t \to 0^+$. It follows that the Mellin transform $MF(s)$ is defined for all $s \in \mathbf{C}$ and is an entire function. This easily implies that $M[J(t) - c_0]$ extends from Re s large to a meromorphic function of $s \in \mathbf{C}$. Taking $M(e^{t\lambda}) = \Gamma(s)\lambda^{-s}$ and using the functional equation $\Gamma(s)\Gamma(1-s) = \frac{\pi}{\sin \pi s}$ gives

$$\frac{\pi}{\sin \pi s} \sum_j (-1)^{j+1} j \varsigma_j(s) = \Gamma(1-s) M[J(t) - c_0]$$

Now take Re $s < 0$ and use the integration formulas

$$\Gamma(1-s)t^{s-1} = \int_0^\infty x^{-2s} e^{-x^2 t} 2x\, dx$$

$$\int_0^\infty e^{-x^2t}\cdot 2x\cdot(4\pi t)^{-\frac{1}{2}}e^{-\ell^2/4t}dt = e^{-\ell x}\ (\ell>0, x>0)$$

to compute $MF(s)$, yielding

$$\Gamma(1-s)M[J(t)-c_0] = \int_0^\infty x^{-2s}\frac{Z'(x)}{Z(x)}dx$$

Write $\frac{Z'(x)}{Z(x)} = E(x) + \frac{2c_0}{x}\chi_{(0,1)}$ for $x>0$, so $E(x)$ is bounded. Then the integral becomes

$$-\frac{c_0}{s} + \int_0^\infty E(x)dx + o(s)$$

Using $\frac{\pi}{\sin\pi s} = \frac{1}{s} + 0(s)$, we have shown

$$-\log T_\rho^2 = \sum(-1)^{j+1}j\zeta_j'(0) = \int_0^\infty E(x)dx$$

But this shows that for $x>0$

$$\begin{aligned} Z(x) &= \exp-\int_x^\infty \frac{Z'(t)}{Z(t)}dt = \exp-\int_x^\infty E\cdot\exp\int_x^1\frac{2c_0}{t}dt \\ &= (T_\rho^2 + o(x))x^{2c_0} \end{aligned}$$

In particular, if ρ is acyclic then $c_0 = 0$ and

$$Z(0) = T_\rho^2$$

Using the Cheeger/Müller theorem, this shows

$$\Psi_E = \tau(E_\rho)^2$$

Using (V) of Section 1 on the sphere bundle $S^{n-1}\to SX\to X$ gives

$$\tau(E) = \tau(E_\rho)\cdot\tau(E_{\hat\rho})^{(-1)^{n-1}}$$

where $\hat\rho = \rho\otimes w_1$ and $w_1\colon \pi_1X\to\{\pm1\}\subset U(1)$ is the first Stiefel-Whitney class ($w_1 = +1$ on orientation preserving loops in X, -1 on orientation reversing loops). Using Milnor's duality theorem for Reidemeister torsion [M4]

$$\tau(E_{\hat\rho})^{(-1)^{n-1}} = \tau(E_{\rho^*})$$

where ρ^* is the contragredient representation of ρ: in the present case ρ is unitary and so $\rho^* = \rho$,

$$\tau(E) = \tau(E_\rho)^2 = \Psi_E$$

and E is Lefschetz. (Note that we have in fact shown $\tau(E) = T_\rho^2$ for any Riemannian X and acyclic unitary ρ).

The partial fraction decomposition of $a + \frac{Z'}{Z}$ shows it is an odd function. Integrating gives the functional equation

$$(F.E.) \qquad e^{2as}Z(s) = Z(-s)$$

Likewise this decomposition and the Lemma imply the product expansion

$$(P.E.) \qquad Z(s) = \tau(E_\rho)^2 e^{-as} \prod_{\lambda>0} \left(1 + \frac{s^2}{\lambda^2}\right)^{c_\lambda} \cdot s^{2c_0}$$

Now *formally* $\tau(E_\rho) = \prod_{\lambda>0} \lambda^{c_\lambda}$ so that the function $e^{as}Z(s)$ is "monic." This curiosity can be explained by evaluating the determinants of the operators $\Delta_j + s^2$. One finds that when these operators are invertible

$$e^{as}Z(s) = \prod_j (\mathrm{Det}(\Delta_j + s^2))^{(-1)^{j+1}j}$$

i.e. $e^{as}Z(s)$ is a ratio of regularized characteristic polynomials, each of which is formally monic.

We now prove Lemma 6.4. By Bieberbach's theorem $X = \mathbf{R}^n/\pi$ where π is a group of isometries that meets the translation group $\mathbf{R}^n$ in a lattice Γ. We view a p-form in E_ρ as a $\mathbf{C}^m$-valued π-equivariant p-form on $\mathbf{R}^n$ and first examine the condition of Γ-equivariance. The restriction $\rho|\Gamma$ decomposes as a sum of weight spaces $V_\alpha \subset \mathbf{C}^m$ where $\alpha \in \hat{\Gamma} = \mathrm{Hom}(\Gamma, S^1)$, $V_\alpha = \{v \in \mathbf{C}^m | \rho(\gamma)v = \alpha(\gamma)v,\ \gamma \in \Gamma\}$. We write $\alpha = e^{2\pi i\beta}|\Gamma$ where $\beta \in (\mathbf{R}^n)^*$-such β form a coset L_α of the lattice Hom $(\Gamma, \mathbf{Z}) = L \subset (\mathbf{R}^n)^*$.

The quotient group $F = \pi/\Gamma$ is a finite subgroup of $0(n)$. Its natural action on $(\mathbf{R}^n)^*$ preserves $\bigcup_\alpha L_\alpha$. To each orbit $\mathcal{O}$ of F on this union we will associate a space $W_{\mathcal{O}}^p$ of p-forms in E_ρ of bounded dimension on which $\Delta_p = ||\beta||^2$ (note this norm is independent of the choice of $\beta \in \mathcal{O}$) such that the spectrum of Δ_p is exhausted by these subspaces.

Observe that Γ-equivariant functions $f: \mathbf{R}^n \to \mathbf{C}^m$ transform under the action of π that sends f to $\rho(g)^{-1} \circ f \circ g$, $g \in \pi$. If $\beta \in L_\alpha$, $v \in V_\alpha$ then $f = e^{2\pi i\beta}v$ is Γ-invariant and satisfies

$$\rho(g)^{-1} \circ f \circ g = \lambda \cdot f' \ , \ |\lambda| = 1,\ f' = \beta' \cdot v'$$

where $\beta' = \beta \circ g$, $v' = \rho(g)^{-1}v$, $\lambda = e^{2\pi i\beta(t_g)}$ and $\lambda_g \in F$ and $t_g \in \mathbf{R}^n$ denote the linear and translational parts of g respectively:

$$gx = \lambda_g x + t_g \ , \ x \in \mathbf{R}^n$$

If $\alpha' = \alpha \circ \lambda_g$ then $\beta' \in L_{\alpha'}$, $v' \in V_{\alpha'}$. Thus the span of these functions $f = e^{2\pi i\beta}v$, $\beta \in \mathcal{O}$ is a π-invariant space $Y_{\mathcal{O}}$. Take the corresponding π-invariant space of p-forms $Y_{\mathcal{O}} \otimes \Lambda^p(\mathbf{R}^n)^* = Y^p_{\mathcal{O}}$ and let $W^p_{\mathcal{O}} = (Y^p_{\mathcal{O}})^F$ be the subspace of $Y^p_{\mathcal{O}}$ fixed by the action of F.

Clearly $\Delta_p|Y^p_{\mathcal{O}} = \|\beta\|^2$. Moreover every Γ-invariant eigenform is in the span of the $Y^p_{\mathcal{O}}$, so every π-invariant eigenform is in the span of the $W^p_{\mathcal{O}}$. Further dim $(W^p_{\mathcal{O}}) \le \binom{n}{p}m(\#F)$ for all $\mathcal{O}$.

We will show that for many $\mathcal{O}$,

$$(*) \qquad \sum(-1)^p p \ \dim W^p_{\mathcal{O}} = 0$$

Indeed for $\beta \in \mathcal{O}$, let $I_\beta \subset F$ be the isotropy group of elements that fix $\beta \in (\mathbf{R}^n)^*$. Suppose I_β fixes a subspace of $(\mathbf{R}^n)^*$ of dimension greater than 1. Assume, say, that I_β fixes dx_1 and dx_2. Take the span $U^p_{\mathcal{O}} \subset Y^p_{\mathcal{O}}$ of the forms $(e^{2\pi i\beta} \cdot v)\ dx^{j_1} \wedge \cdots \wedge dx^{j_p}$, $3 \le j_1 < \ldots < j_p \le n$, $v \in V_\alpha$. Let $F^p_{\mathcal{O}} = (U^p_{\mathcal{O}})^{I_\beta}$ be the I_β-fixed subspace of $U^p_{\mathcal{O}}$. By assumption $F^p_{\mathcal{O}} \cdot \omega = (U^p_{\mathcal{O}} \cdot \omega)^{I_\beta}$ for $\omega = 1, dx^1, dx^2, dx^1 \wedge dx^2$. Taking an element of $F^p_{\mathcal{O}} \cdot \omega$ and summing over its images under F/I_β gives the general element of $W^{p+|\omega|}_{\mathcal{O}}$ whose β-component involves exactly p of $dx^3, \ldots, dx^n$. Thus

$$\dim W^p_{\mathcal{O}} = \dim F^p_{\mathcal{O}} + 2\dim F^{p-1}_{\mathcal{O}} + \dim F^{p-2}_{\mathcal{O}}$$

and so $(*)$ holds. But $(*)$ implies that the $W^p_{\mathcal{O}}$ make a net contribution of zero to the c_λ's.

The sum $\sum\limits_{\lambda>0} \frac{|c_\lambda|}{\lambda}$ thus merely has contributions from nonzero β for which I_β fixes $\mathbf{R}\beta$ only. As F is finite, there are finitely many choices for I_β and so finitely many lines in $(\mathbf{R}^n)^*$ contain all such β's. Each such line meets $\bigcup\limits_\alpha L_\alpha$ in a uniformly discrete subset. As $\lambda = \|\beta\|^2$ and $\Sigma \frac{1}{n^2} < \infty$ one has

$$\sum_{\substack{\lambda>0 \\ c_\lambda \neq 0}} \frac{1}{\lambda} < \infty$$

Since each line meets the sphere of radius $\lambda^{1/2}$ only twice and each intersection point contributes only a bounded amount to c_λ, we have that $|c_\lambda|$ is uniformly bounded. This proves Lemma 6.4.

It is interesting to consider the case X =Klein bottle. Then $\pi = \langle x, y | x^2 = y^2 \rangle$ where x, y are each glide reflections in the plane. Viewing X as the suspension of an orientation reversing isometry $r: S^1 \to S^1$, x, y correspond to closed orbits γ_x, γ_y swept out by the fixed points of r. By perturbing r to a north pole-south pole diffeomorphism, one finds a Morse-Smale perturbation of this isometric flow with only γ_x, γ_y as closed orbits. This gives

$$\tau(E) = \frac{|\det I + \rho(x)|}{|\det I - \rho(y)|} = \frac{|\det I + \rho(y)|}{|\det I - \rho(x)|}, \ \rho = \rho_E$$

according to whether γ_x is a source and γ_y is a sink or vice versa, provided the indicated determinants are nonzero (see §3). The equality of the two expressions for τ follows from $x^2 = y^2 \in \pi$ or from Milnor duality. The isolated closed geodesics γ are just the odd powers of $\gamma_x, \gamma_y, \gamma_x^{-1}, \gamma_y^{-1}$: as Sunada and Urakawa showed, only finitely many prime closed geodesics are isolated. The group F is of order 2 and it acts on $(\mathbf{R}^n)^*$ by a reflection. The fixed set of this reflection is the only line that is cut out by its isotropy group, hence the only line that makes a nontrivial contribution to the c_λ's and to Z. One can explicitly calculate Z here: we leave this as an exercise.

In the simplest case $X = \mathbf{R}/2\pi\mathbf{Z}$

$$Z(s) = \det(I - \rho(t)e^{-2\pi s}) \det(I - \rho(t^{-1})e^{-2\pi s})$$

where t, t^{-1} are the 2 generators of $\pi_1 X$.

We now discuss the case of constant negative curvature. By more complicated applications of the Selberg trace formula one can show

Theorem 6.5 *For oriented X of constant negative curvature, any acyclic unitary E over SX is Lefschetz*

The proof goes differently according to whether $n = \dim X$ is 2 or greater. For $n > 2$, $\pi_1(SX) \simeq \pi_1 X$ and E necessarily pulls back from a bundle on X. Thus E acyclic $\Rightarrow$ $\chi(X) = 0 \Rightarrow n$ odd. This case was treated in [F3] using Selberg- and Ruelle-type zeta functions and the Cheeger/Müller theorem. For $n = 2$ the homotopy sequence of the fibration $SX \to X$ gives

$$1 \to \mathbf{Z} \to \pi_1(SX) \to \pi_1 X \to 1$$

and one has many acyclic E, even though $\chi(X) < 0$. This case was treated in [F4] (where we even allowed elements of finite order in our discrete group). We have

$$Z_E(s) = \left(\prod_\gamma \det I - \rho(\gamma)e^{-s\ell(\gamma)}\right)^{(-1)^{n-1}}$$

because each closed orbit is hyperbolic with unstable dimension $n-1$ and the unstable manifold oriented (recall X is oriented here). For $n=2$ we found for $t \in \pi_1(SX)$ the class of the circle fiber (i.e. the generator of $\mathbf{Z}$)

$$|\Psi_E| = |\det I - \rho(t)|^{2-2g}$$

which, by the formula of §1 for circle bundles, is just $\tau(E)$. (When elements of finite order are present, one has a *locally* free S^1 action and the exceptional fibers contribute to both Ψ_E and $\tau(E)$ [F4]). In all cases an important role is played by the functional equations of Selberg zeta functions, which is analogous to the functional equation proved above for the torsion function of a flat manifold. We will show in a further paper that our proof of 6.5 holds for nonoriented X as well.

Altogether, by a mixture of analytic and topological methods, any acyclic unitary E is Lefschetz when X has constant curvature.

We end with a remark about the hypothesis that E is unitary. It seems likely that the Cheeger/Müller theorem remains true for a nonunitary acyclic E in the form

$$\tau(E \oplus E^*) = T_{\rho\oplus\rho^*}$$

If this is so then all the results of this section should be true for nonunitary E, just as we have shown for positive curvature. It would be necessary to adapt the results on Selberg zeta functions to nonunitary E, but this has already been done in some cases. Ruelle's method can be used to show that the relevant Selberg zeta functions continue to meromorphic functions on $\mathbf{C}$ and for many E the functional equations can be proved by analytic continuation from the unitary representations [F5,F6].

Assuming that all acyclic bundles over a hyperbolic X are Lefschetz, one can show that (certain of) Tomter's G-induced Anosov flows [T1,T2] are Lefschetz. Such a flow is defined on a flat T^n-bundle Y over the unit sphere bundle SX of a hyperbolic X and the flow on Y corresponds to geodesic flow on SX. Both the Fuller measure and R-torsion are functorial by (IV) of §1, §4. So for a bundle E on SX, the pulled back bundle E' on Y

will be acyclic and Lefschetz provided the bundles $E \otimes H^i(T^n)$ are acyclic and Lefschetz. The bundles $H^i(T^n)$ over SX that Tomter uses are certainly not unitary for $0 < i < n$, so the presently known results do not suffice.

§7. Geometric 3-manifolds

A manifold X has a geometry modelled on a space (M, G), where M is a connected manifold with an effective transitive action of a Lie group G, if there exists a family of charts taking open sets in X to open sets in M and such that the overlap maps are restrictions of actions of elements of G. For G a group of isometries of a Riemannian metric on M, X inherits a metric so that the charts are isometries. Thurston studied the case when dim $M = 3$, M is complete, G is unimodular, and no larger group of homeomorphisms of M has compact point stabilizers: he found that only 8 pairs (M, G) were needed to describe geometries on connected closed 3-manifolds X [Th]. We will show that for such a geometric 3-manifold X, many acyclic bundles are Lefschetz.

Of the 8 geometries in question, 3 are of constant curvature, and so have been treated in §6. Another 2 geometries arise by taking as model space $\mathbf{R} \times M_\epsilon^2$ where M_ϵ^2 denotes the hyperbolic plane or the 2-sphere for $\epsilon = -1$ or $+1$ respectively. Still another 2 geometries arise by taking twisted products of $\mathbf{R}$ over the Euclidean and hyperbolic plane, giving respectively the Heisenberg group N and the universal cover of the sphere bundle of $M_{-1}^2, \widetilde{SM}_{-1}^2$ as model space. We will see that these 4 geometries can be treated much like the spherical geometry studied in §6.

For any one of these 4 model spaces, G preserves the $\mathbf{R}$ fibers and acts on them by isometries. Let $G^+ \subset G$ denote the subgroup of G that preserves the orientation of these $\mathbf{R}$ fibers. Then for any X modelled on (M, G) there is a connected cover X^+ with at most $[G: G^+] \leq 2$ sheets such that X^+ is modelled on (M, G^+). Consider a unit vector tangent to an $\mathbf{R}$-fiber and extend it to a G^+-equivariant vector field Y on M whose flow ψ defines a one parameter subgroup of G^+. Then X^+ inherits a vector field Y^* whose flow $\phi_t: X^+ \to X^+$ is isometric. Thus X^+ has an isometry group $G(X^+)$ of positive dimension, hence an isometric action of S^1. Choose the generator of this S^1 action near Y^* so this action is locally free. Since geodesic flow commutes with isometries, this puts us in the setting of §5: Theorem 5.1 assures us that bundles over X^+ whose holonomy around a circle fiber fixes no vector are acyclic and Lefschetz. Inducing from X^+ to X gives Lefschetz bundles on X, by the naturality property (VI) of §1, §4.

Now we turn to the eighth geometry, where M is $\mathbf{R}^3$ with the metric such that the usual unit vectors at $(x,y,z) \in \mathbf{R}^3$ are orthogonal with $\|\frac{\partial}{\partial x}\| = e^z$, $\|\frac{\partial}{\partial y}\| = e^{-z}$, $\|\frac{\partial}{\partial z}\| = 1$. Then translations in the $x-y$ plane generate a subgroup T of isometries that is normalized by the one parameter group $(x,y,z) \to (e^t x, e^{-t} y, z+t)$. The resulting semidirect product $G^o = T \rtimes \mathbf{R}$ is of index 4 in G, with the stabilizer of the origin generated by the involutions $\alpha: (x,y,z) \to (-x,-y,z)$ and $\beta: (x,y,z) \mapsto (y,x,-z)$. Let Y be the vector field $\frac{\partial}{\partial z}$ on M and $G^+ = G^o \rtimes \{1,\alpha\}$ its stabilizer. Note that the flow ψ generated by Y does not consist of isometries in this case. Still with X^+ as before, Y induces a vector field Y^* and a flow $\phi_t: X^+ \to X^+$. Moreover since G^+ preserves the 1-form dz, X^+ inherits a closed 1-form ω with $\omega(Y^*) = 1$. Thus ϕ_t has a cross-section: indeed one can see that $H^1(X^+;\mathbf{R})$ is 1-dimensional, generated by the class of ω, so X^+ fibers over S^1 by the map

$$\theta(X) = \int_{x_0}^{x} \omega$$

where the integral is taken over any path from a basepoint x_0 to x. The ambiguity in the path determines an ambiguity in θ of some period of ω, but such periods form an infinite cyclic subgroup of $\mathbf{R}$ and so θ takes values in the quotient S^1. One sees that X^+ is a flat T^2-bundle over S^1 with monodromy a linear map A with eigenvalues λ, λ^{-1}, $\lambda \in \mathbf{R}, |\lambda| > 1$: conversely any such bundle can be modelled on (M, G^+).

The geodesic flow on X^+ can be studied using the fact that here M is isometric to the Lie group G^o with a left-invariant metric. This allows one to set up Euler equations on S^2 whose integral curves describe the velocity vector of a geodesic relative to the G^o-invariant orthonormal frame $e^{-z}\frac{\partial}{\partial x}$, $e^{z}\frac{\partial}{\partial y}$, $\frac{\partial}{\partial z}$. This was done by Grayson who found in this way that the only closed geodesics γ on X^+ with $\int_\gamma \omega > 0$ were flowlines of ϕ_t [G]. Such γ clearly contribute to the Fuller measure μ for the geodesic flow on SX^+ enough to contribute $\tau(E)$ to $\Psi_{\phi,E}$, for any acyclic E over X^+, by the results of §3 for circular flows. Reversing such γ's contributes another factor of $\tau(E)$. Since X is oriented the bundle E^* on SX corresponding to E on X satisfies

$$\tau(E^*) = \tau(E)^2$$

by (IV) of §1. We will have that E^* is Lefschetz provided we show that γ's with $\int_\gamma \omega = 0$ contribute zero to μ.

To see this we first refer to [G] to see that each γ with $\int_\gamma \omega = 0$ lies in a closed submanifold of geodesics that we denote S. S is determined by fixing a value of θ and a point p on S^2 that's periodic for the Euler flow, taking all the unit vectors based on the fiber over θ with direction p and letting these unit vectors flow for a period t divisible by the period of ϕ for the Euler flow. It is then clear that one can compute the Fuller index of S by first passing to the infinite cyclic cover $\tilde{X}$ on which θ is single-valued. One has $\pi_1\tilde{X} = \pi_1 X^+ \cap T$ and the quotient 2-torus $T/\pi_1\tilde{X}$ acts freely and isometrically on $\tilde{X}$. This shows that S has Fuller index 0: as in Lemma 5.3 the essential point is that the orbit space of S carries a free S^1 action and so has Euler characteristic zero.

Again we pass from X^+ to X by induction to produce Lefschetz bundles over X. Thus we have shown that for all 8 geometries a large class of acyclic bundles over a 3-manifold with such a geometry are Lefschetz. In the 5 cases of nonconstant curvature if a certain line field is orientable then any "reasonable" acyclic E is Lefschetz.

In conclusion, the results of the last two sections are evidence for the following

Conjecture If X is a compact locally homogeneous Riemannian manifold and E is an acyclic bundle over SX then E is Lefschetz.

It is even conceivable that (ϕ, E) is Lefschetz for any acyclic E with a flat density and any C^ω contact flow ϕ. The Fuller index seems more tractable for contact flows and they are conjectured to always have closed orbits. This would of course imply that C^ω geodesic flows are Lefschetz.

BIBLIOGRAPHY

[AB] Atiyah, M. and R. Bott. A Lefschetz fixed point formula for elliptic complexes, II. *Annals of Math.* **88** (1968) 451-491.

[B] Birkhoff, D. Dynamical systems with two degrees of freedom. *Trans. AMS* **18** (1917) 199-300.

[C] Cohen, M. A course in simple-homotopy theory. *Springer Verlag GTM* **10**, 1973.

[CE] Conley, C. and R. Easton. Isolated invariant sets and isolating blocks. *Trans. AMS* **158** (1971) 35-61.

[Ch] Cheeger, J. Analytic torsion and the heat equation. *Annals of Math.* **109** (1979) 259-322.

[CMP] Chow, S. and J. Mallet-Paret. The Fuller index and global Hopf bifurcation. *J. Diff. Eq.* **29** (1978) 66-85.

[Co] Conley, C. Isolated invariant sets and the Morse index. *CBMS Reg. Conf.* **38**, 1978.

[DR] De Rham, G. Sur les complexes avec automorphismes. *Comment. Math. Helv.* **12** (1939) 191-211.

[F1] Fried, D. Homological identities for closed orbits. *Inv. Math.* **71** (1983) 419-442.

[F2] Fried, D. The geometry of cross-sections to flows. *Topology* **21** (1982) 353-371.

[F3] Fried, D. Analytic torsion and closed geodesics on hyperbolic manifolds. *Inv. Math.* **84** (1986) 523-540.

[F4] Fried, D. Fuchsian groups and Reidemeister torsion. in *The Selberg trace formula and related topics, Contemp. Math.* **53**, 1986.

[F5] Fried, D. Zeta functions of Ruelle and Selberg, I. To appear in *Ann. Ec. Norm. Sup.*

[F6] Fried, D. Zeta functions of Ruelle and Selberg, II. In preparation.

[Fr1] Franks, J. Knots, links and symbolic dynamics. *Annals of Math.* **113** (1981) 529-552.

[Fr2] Franks, J. Homology and dynamical systems. *CBMS Reg. Conf.* **49**, 1982.

[Fra] Franz, W. Über die Torsion einer Überdeckung. *J. Reine Angew. Math.* **173** (1935) 245-254.

[Fu1] Fuller, F.B. On the surface of section and periodic trajectories. *Amer. J. Math.* **87** (1965) 473-480.

[Fu2] Fuller, F.B. An index of fixed point type for periodic orbits. *Amer. J. Math.* **89** (1967) 133-148.

[G] Grayson, M. Geometry and growth in three dimensions. Ph.D. thesis, Princeton Univ., 1983.

[GS] Guillemin, V. and S. Sternberg. Geometric asymptotics. *AMS Math. Surveys* **14**, 1977.

[L] Lefschetz, S. Continuous transformations of maniflds. *Proc. Nat. Acad. Sci. USA* **9** (1923) 90-93.

[La] Lang, S. *Algebra*. Addison-Wesley, 1965.

[M1] Milnor, J. Infinite cyclic coverings. In *Conf. on the Topology of Mflds,* Hocking, ed. Prindle, Weber, Schmidt, 1968.

[M2] Milnor, J. Topology from a differentiable viewpoint. Univ. of Va., 1966.

[M3] Milnor, J. Whitehead torsion. *Bull. AMS* **72** (1966) 358-426.

[M4] Milnor, J. A duality theorem for Reidemeister torsion. *Annals of Math.* **76** (1962) 137-147.

[Mo] Montgomery, J. Cohomology of isolated invariant sets under perturbation. *J. Diff. Eq.* **13** (1973) 257-299.

[Mu] Müller, W. Analytic torsion and R-torsion of Riemannian manifolds. *Adv. in Math.* **28** (1978) 233-305.

[NC] de Neymet de Christ, S. Some relations in Whitehead torsion (in Spanish), *Bull. Soc. Mat. Mexicana* **12** (1967) 55-70.

[R] Reidemeister, K. Homotopieringe und Linsenräume. *Hamb. Abh.* **11** (1935) 102-109.

[RS] Ray, D. and I. Singer. R-torsion and the Laplacian on Riemannian manifolds. *Adv. in Math.* **7** (1971) 145-210.

[Sc] Schwartzman, S. Asymptotic cycles. *Annals of Math.* **66** (1957) 270-284.

[Sch] Schweitzer, P. Counterexamples to the Seifert conjecture and opening closed leaves of foliations. *Annals of Math.* **100** (1971) 386-400.

[So] Soucek, J. Morse-Sard theorem for closed geodesics. *Comment. Math. Univ. Carolin.* **25** (1984) 265-272.

[SU] Sunada, T. and H. Urakawa. Ray-Singer zeta functions for compact flat manifolds, preprint.

[T1] Tomter, P. Anosov flows on infra-homogeneous spaces. In *Proc. Symp. Pure Math XIV*, Providence, 1970.

[T2] Tomter, P. On the classification of Anosov flows. *Topology* **14** (1975) 179-189.

[Th] Thurston, W. Three dimensional manifolds, Kleinian groups and hyperbolic geometry. *Bull. AMS* **6** (1982) 357-381.

[Wa] Watts, C. On the Euler characteristic of polyhedra. *Proc. AMS* **13** (1962) 304-306.

[W] Wilson, F.W. On the minimal sets of non-singular vector fields. *Annals of Math.* **84** (1966) 529-536.

Mathematics Department
Boston University
111 Cummington St.
Boston, MA 02215
USA

Contemporary Mathematics
Volume **58**, Part III, 1987

PHASE LOCKING AND HIGHER ORDER AVERAGING

Catherine Garcia
A. A. Minzoni

ABSTRACT. In this work we obtain explicit expressions for the phase locking regions for a differential equation with periodic coefficients with zero first order averaging. This completes the work of Hoppensteadt who computed the intervals for equations with non-zero first order averaging. This process provides a straight forward computational approach to obtain the qualitative results for the phase locking regions for circle mappings discussed in [1]. In the second part the same ideas are used for phase Locking in a two dimensional, time dependent, Hamiltonian system. The standard implicit function theorem or Lyapunov type of arguments are replaced by the Moser twist theorem to ensure phase locking

INTRODUCTION. The first three sections of the paper are concerned with the computation of the phase locking intervals for a differential equation which describes a Voltage controlled oscillator. The calculations are performed at a formal level using the method of multiple scales and justified later using classical results in differential equations. This computational approach reproduces and gives explicit expressions for the qualitative results of [1].

The following sections are concerned with a Hamiltonian example. The difficulties in the situation is the lack of stable orbits which is a sufficient condition in known examples for phase locking. However for 2 dimensional systems with time dependence it is possible to prove the boundedness of the relevant solutions using the results of Moser on invariant curves. An example is discussed in great detail and it is indicated how to proceed for more general Hamiltonians. A speculation is given for higher dimensional Hamiltonian systems.

FORMULATION OF THE PROBLEM AND CUALITATIVE RESULTS. We study the phase locking properties for a simple ordinary differential equation for the phase of

© 1987 American Mathematical Society
0271-4132/87 $1.00 + $.25 per page

a voltage controlled oscillator. The equation to be studied is

$$\dot{\theta} = w + \epsilon \operatorname{Cos} \theta(t) + \epsilon \operatorname{Cos} \nu t$$

Where W is the intrinsic frequency of the oscillator and ν the frequency of the imput signal. The parameter ϵ will be assumed small. Changing variables and calling νt again t we have:

$$\text{(1)} \qquad \dot{\theta} = \frac{W}{\nu} + \frac{\epsilon}{\nu} (\cos \theta(t) + \operatorname{Cos} t).$$

We are interested in studyng the behavior of the solutions for fixed ν as functions of w. In particular we want to find the value of w for which there are stable periodic orbits of equation (1) on the torus $0 \leqslant \theta \leqslant 2\Pi$, $0 \leqslant t \leqslant 2\Pi$. These periodic orbits are the "phase locked" solutions. This is to say solutions in θ with period which is a rational multiple of the period of the forcing function. When those solutions are present suitable initial values will generate solutions which approach the phase locked ones.

Equation (1) is an example but with the same methods we can study equations of the form

$$\text{(2)} \qquad \dot{\theta} = w + \epsilon\, p(\theta,t)$$

with P a periodic function of the two variables.

The qualitative theory for (1) and (2) is well established and we just recall the results.

Consider the Poincaré map associated with (1). Defined by $T(\epsilon,w)\theta_0 = \theta(\epsilon,w,2\Pi)$ where $\theta(\epsilon,w,2\Pi)$ in the solution of (1) with initial value θ_0. The mapping T is θ diffeomorphism of the circle.

Fix a value of ϵ say ϵ_0. By the results of Arnold [1] it follows that there are open regions of the parameter w in which the rotarion number of T is rational. In those regions there are periodic orbits, the fixed points of T alway even in number, half of them stable and half of them unstable. From the qualitative standpoint all the behavior in known. However this theory gives no information on the size and location of the regions of phase locking and the corresponding rotation number.

In this work we are interested in computing explicity, for small ϵ, the phase loking intervals. For this purpose we use the method of averaging which has been used to first order in this context by Hoppensteadt.

The idea is to compute the rotation number from the solutions of the equation using the fact that rotation number ρ is given by

$$\rho(\epsilon,w) = \lim_{t\to\infty} \theta(t,e,w,\theta_0)/t$$

where $\theta(t,\epsilon,w,\theta_0)$ is the solution of (1) with initial value θ_0. If this rotation number (independent of θ_0) in shown to be national for θ set of values of w and ϵ we will have the desired phase locking region. To compute the rotation number it is necessary to find the behavior of the solutions as $t\to\infty$. This is accomplhished in [5] using the method of averaging. In the next section we use higher order averaging to compute the phase locking regions not accesible using the methods of [5].

HIGHER ORDER AVERAGING AND LIMITING BEHAVIOR AS $t\to\infty$. Let $\theta = \frac{wt}{\nu} + \varphi$ the equation for φ becomes

$$\dot{\varphi} = \frac{\epsilon}{\nu} \operatorname{Cos} \left(\frac{w}{\nu} t+\varphi\right) + \frac{\epsilon}{\nu} \operatorname{Cos} t \tag{3}$$

In [5] equation (3) is averaged and one shows that φ is bounded for all $t \geqslant 0$ and is easy to compute the rotation number. However (3) has zero average for $\frac{w}{\nu}$ not small.

Therefore one must go to the next order. One could attemp to make the changes of variables involved in the averaging [5] recursively to reduce the equation to the average plus an oscillatory remainder. This will be done in the next section to prove the bounds on φ. At this stage we proceed formally as in [6] since this procedure in more convenient computationally and provide the proofs of the validity in the next section.

We thus let

$$\varphi = \xi + \epsilon v_1(\xi,t) + \epsilon^2 v_2(\xi,t) + \epsilon^3 v_3(\xi,t) + \ldots..$$

$$\dot{\xi} = \epsilon\, M(\xi) + \epsilon^2 M_1(\xi) + \epsilon^3 M_3(\xi) + \ldots..$$

substitute into the equation, equate powers of ϵ and use the requirement of periodicity in the function σ_i to find the average, autonomous equation, for ξ. We obtain after some algebra

$$O(\epsilon)\quad v_{1_t}(\xi,t) + M_1(\xi) = \frac{1}{\nu} \{\operatorname{Cos} (\tfrac{w}{\nu} t+\xi) + \operatorname{Cos} t\}$$

$$O(\epsilon^2)\quad v_{2_t}(\xi,t) + M_2(\xi) = \frac{1}{\nu} \{\tfrac{1}{2} - \tfrac{1}{2} \operatorname{Cos} (\tfrac{2w}{\nu} t+2\xi)\}$$

$$+ \frac{1}{\nu} \{\tfrac{1}{2} \operatorname{Cos} ((\tfrac{w}{\nu} - 1)t+\xi) - \tfrac{1}{2} \operatorname{Cos} ((\tfrac{w}{\nu}+1)t+\xi)\}$$

Observe that we must take $M_1 = 0$ to obtain v_1 as a bouded function of t. To obtain a bouded solution for v_2 two cases are possible. In the first instance $\frac{w}{\nu} - 1 = O(1)$, in this case $M_2 = -\frac{1}{2\nu}$ and no information in obtained to this order. The more interesting case is for $\frac{w}{\nu} - 1 = \epsilon^2\sigma$. In this case we must take M in the form

$$M_2 = -\frac{1}{\nu}\{\frac{1}{2} + \frac{1}{2} \operatorname{Cos}(\epsilon^2\sigma t+\xi)\}$$

To this order the equation for ξ becomes

$$\dot{\xi} = -\frac{\epsilon^2}{\nu}(\frac{1}{2} + \frac{1}{2}\operatorname{Cos}(\epsilon^2\sigma t+\xi)) \tag{3b}$$

Let now $\xi = \eta - \epsilon^2\sigma t$ to obtain an equation for η in the form:

$$\dot{\eta} = \epsilon^2(\sigma - \frac{1}{2\nu} - \frac{1}{2\nu}\operatorname{Cos}\eta) \tag{4}$$

Equation (4) has two critical points provided $\left|\sigma - \frac{1}{2\nu}\right| < \frac{1}{2\nu}$. One of the critical points is stable, the other unstable. Therefore to this order we have that for $0 \leqslant \sigma \leqslant \frac{1}{2\nu}$ the solution for θ takes the form

$$\begin{aligned} \theta(t) &= \frac{w}{\nu}t + \varphi = \frac{w}{\nu}t + \xi\epsilon v_1 + \dots \\ &= \frac{w}{\nu}t - \frac{w}{\nu}t + t + \eta + \epsilon v_1 \dots = t + \eta + \epsilon v_1 + \dots\dots \end{aligned} \tag{5}$$

If we can prove that the remainder $\epsilon v_1 + \dots$ is bounded for all times it follows from (5) that for $0 \leqslant \frac{w}{\nu} - 1 \leqslant \frac{\epsilon^2}{2\nu}$ the rotation number ρ is given by:

$$\rho = \lim_{t\to\infty} \frac{\theta(t)}{t} = 1$$

and that the stable periodic solution is given by $t+\eta_s$ where η_s in the stable critical point of (4). This locates the first locking region. The width is $O(\epsilon^2)$. The next region is obtained averaging to higher order when we assume $\frac{w}{\nu} - 1$ to be of order one. If one carries the procedure to the next order one obtains for the average equation assuming now $\frac{2w}{\nu} - 1$ small

$$\dot{\xi} = -\frac{\epsilon^2}{2\nu} - \frac{\epsilon^3}{4(w-\nu)}\operatorname{Cos}(\frac{2w-\nu}{\nu}t + 2\xi)$$

Let now $\xi = -\frac{1}{2}(\frac{2w-\nu}{\nu})t + \eta$ and let $\frac{2w-\nu}{\nu} = \epsilon^2\sigma$ to obtain:

$$\dot{\eta} = \frac{\epsilon^2}{2}\sigma - \frac{\epsilon^2}{2\nu} - \frac{\epsilon^3}{4(w-\nu)}\operatorname{Cos} 2\eta \tag{6}$$

Therefore for $|\sigma - {}^1/_\nu| \leqslant \mu\epsilon$ for some $\mu > 0$ there are 4 critical points for equation (6). Two stable and two unstable. They correspond, to this order to the periodic phase locked orbits.

Again it follows from (6) that for

$\left|\frac{2w-\nu}{\nu} - \frac{2}{\nu}\right| \leqslant \mu\epsilon^3$ the rotation number is $\frac{1}{2}$ in accordance with the general qualitative results of [1].

To justify this procedure in necessary to show the boundedness of the remainders and the existence for the desired periodic solutions. This is proved in the next section for the averaging $O(\epsilon^2)$.

It is important to notice that this procedure is not uniform in ϵ, this is to say that higher order resonances may be captured by this method only for ϵ smaller than lower order resonances. Also the width of the regions become very small the higher the resonance order.

PROOF OF THE VALIDITY OF THE AVERAGING. In the equation for φ introduce the exact change of variables

$$\varphi = \xi + \epsilon\, v_1(t,\xi) + \epsilon^2 v_2(t,\xi)$$

Where the functions v_1 and v_2 are formally computed using the equations of the previous section. Now obtain an exact equation for $\dot{\xi}$. Because of the construction of v_1 and v_2 to $O(\epsilon^2)$ the only terms which are slowly varying in t are the same ones as the ones in equation (3b). there is now a remainder which comes from the finite Taylor series expansion to cubic order of $\operatorname{Cos}(\frac{W}{\nu}t + +\epsilon v_1 + \epsilon^2 v_2)$ and from the $O(\epsilon)$ terms obtained from the v_{1_ξ} and v_{2_ξ} derivatives. The exact equation for $\dot{\xi}$ takes the form

$$\dot{\xi} = -\frac{\epsilon^2}{\nu}\left(\frac{1}{2} + \frac{1}{2}\operatorname{Cos}(\epsilon^2\sigma t + \xi) + \epsilon^3 F(\operatorname{Sin}(\tfrac{W}{\nu}t+\xi), \operatorname{Sin}((\tfrac{W}{\nu}+1)t+\xi),\right.$$

$$\left.\operatorname{Sin}(2\tfrac{W}{\nu}T+2\xi), \operatorname{Cos}(\tfrac{W}{\nu}t+\xi), \operatorname{Cos}((\tfrac{W}{\nu}+1)t+\xi), \operatorname{Cos}(2\tfrac{W}{\nu}t+2\xi)\right)$$

and $\epsilon^2\sigma = \frac{W}{\nu} - 1$. Taking now $\xi = -\epsilon^2\sigma t + \eta$ we obtain the following exact equation for η

$$(7)\qquad \dot{\eta} = \epsilon^2 \{\sigma - {}^1/{}_{2\nu} - \frac{1}{2\nu}\operatorname{Cos}\eta\} + \epsilon^3 p(t,\eta)$$

Where the function $p(t,\eta)$ is 2Π perioric in t. But equation (7) is in the standard form and the classical results of Hale [4] apply to show the existence for $|\sigma - {}^1/{}_{2\nu}| < {}^1/{}_2$ of a stable 2Π periodic solution η_s and an unstable 2Π periodic solution η_u.

Also for initial conditions sufficiently close to the critical point of the $O(\epsilon^2)$ equation (7) all the solutions converge to the stable periodic solution. In particular the solutions η are bounded for all time and this justifies the formal calculation of the rotarion number assuming boundedness for η to all orders.

PHASE LOCKING FOR A HAMILTONIAN SYSTEM. In this section we extend the phase locking results of [5] to a time dependend two dimensional Hamiltonian system. We work with a special example in the same spirit of [5] which represents coupled voltage controlled oscillators. The method used in this section works for every hamiltonian system whose average has closed orbits with amplitude-dependent period. This is the situation usually encountered in practice.

Let θ_1, θ_2, θ_3, θ_4 be the phases of the V.C.O.S. We assume that oscillator 1 excites 2 and oscillator 2 excites oscillator 1.

Also oscillator 3 inhibits 1 while 4 inhibits 2. For illustration we take the system to be

$$
\begin{aligned}
\dot{\theta}_1 &= w - \epsilon \operatorname{Sin} 2(\theta_1+\theta_2-\theta_3) + \epsilon \operatorname{Cos}(\theta_1-\theta_2) - \epsilon H^{(1)}_{\theta_2}(\theta_1,\theta_2,t) \\
\dot{\theta}_2 &= w + \epsilon \operatorname{Sin} 2(\theta_1+\theta_2-\theta_3) + \epsilon \operatorname{Cos}(\theta_1-\theta_2) - \epsilon H^{(1)}_{\theta_1}(\theta_1,\theta_2,t) \qquad (8) \\
\dot{\theta}_3 &= 2w_0, \quad \dot{\theta}_n = 2w_0
\end{aligned}
$$

The function $H^{(1)}(\theta_1,\theta_2,t)$ is assumed 2π periodic in t with zero average in t. This function accounts, for external stimulation and interactions between oscillators one and two.

By phase locking we mean that for w chose to w_0 appropriately chosen we have $\lim_{t\to\infty} \frac{\theta_1(t)}{\theta_2(t)} = 1$; for all initial conditions $\theta_1(0)$, $\theta_2(0)$ suitably chosen.

In the work of Hoppensteadt and Keener this type result is provided the system has dissipation which implies asymptotic stability of the "phase locked" orbits.

This is impossible for a Hamiltonian system. However the same arguments used in [5] will give the result if it possible to obtain a bound on the solutions. For two dimensional Hamiltonian systems the bound in non-trivial and must be obtained using the results of Moser on in variant curves for area preserving mappings.

We thus look for phase locked solutions for $w_0 = w_0 + \epsilon w_1$ in the form

$\theta_1 = w_0t+\varphi_1$, $\theta_2 = w_0t+\varphi_2$. The equations for φ_1 and φ_2 are

$$\overset{\circ}{\varphi}_1 = \epsilon w_1 - \epsilon \operatorname{Sin} 2(\varphi_1+\varphi_2) + \epsilon \operatorname{Cos}(\varphi_1-\varphi_2) - H^{(1)}_{\varphi_2}(\varphi_1,\varphi_2,t)$$

(9)

$$\overset{\circ}{\varphi}_1 = \epsilon w_1 - \epsilon \operatorname{Sin} 2(\varphi_1+\varphi_2) + \epsilon \operatorname{Cos}(\varphi_1-\varphi_2) + H^{(1)}_{\varphi_1}(\varphi_1,\varphi_2,t)$$

If we prove

$$\lim_{t\to\infty} \frac{\varphi_1(t)}{t} = \lim_{t\to\infty} \frac{\varphi_2(t)}{t} = 0$$

for all w_1 sufficity small provided $\varphi_1(0)$, $\varphi_2(0)$ one appropriately chosen we will have phase locking for equations (8).

To estimate the solutions φ_1, φ_2 we average the equations. However to keep the system Hamiltonian this is accomplished introducing the canonical transformation [2] to the variables Φ_1, Φ_2 using the generating function.

$$S(\Phi_1,\varphi_2) = \varphi_2\Phi_1 + \epsilon \int_0^t H^{(1)}(\Phi_1,\varphi_2,S)\, ds$$

and the new variables defined by:

$$\varphi_1 = S_{\varphi_2} = \Phi_1 + \epsilon \int_0^t H^{(1)}_{\varphi_2}(\Phi_1,\varphi_2,S)ds$$

(10)

$$\Phi_2 = S_{\Phi_1} = \varphi_2 + \epsilon \int_0^t H^{1}_{\Phi_1}(\Phi_1\ \varphi_2,S)ds$$

For ϵ sufficiently small it follows by implicit functions that the transformation defined by equations (10), $\varphi_1 = \varphi_1(\Phi_1,\Phi_2,\epsilon)$, $\varphi_1 = \varphi_2(\Phi_1,\Phi,\epsilon)$ is an isomorphism between any bounded region in the Φ_1,Φ_2 plane and its image in the φ_1,φ_2 plane. Also the transformation are analytic in ϵ.

The new Hamiltonian takes the form

$$\overline{H}(\Phi_1,\Phi_2) = \epsilon\{\operatorname{Sin}^2 \Phi_1+\Phi_2) + \operatorname{Sin}(\Phi_1-\Phi_2) + w_1(\Phi_1,\Phi_2)\}$$
$$+ \epsilon^2 \hat{H}(\Phi_1,\Phi_2,t)$$

Since the canonical transformation (10) was chosen to remove the $O(\epsilon)$ oscillatory therms. The therm $\tilde{H}$ will have a non zero average.

We can now show that $\Phi_1+\Phi_2 = p$, $\Phi_1-\Phi_2 = q$ are bounded for all times.

The $O(\epsilon)$ Hamiltonian in the p,q variables in the pendulum Hamiltonian for $w_1 = 0$. Then there are periodic orbits for small w_1. These orbits one the invariant curves for the Poincaré map associated with the solution of the averaged equations form $t=0$ to $t=\frac{2\Pi}{w_0}$. We will show now using the results on persistence of invariant curves under perturbations that the time dependent problem has also invariant curves. This gives boundedness of p,q for all times on hence for Φ_1,Φ_2 which is what we want to prove.

To show the existence of invariant curves in necessary to introduce action angle variables fot the $O(\epsilon)$ Hamiltonian. This is done in the usual way introducing the generating function [2] $W(q,J)$ defined by

$$W(q,E) = \int_0^q \sin^{-1} \{ E - \sin \xi + w_1 \xi \}^{1/2} d\xi$$

where the energy E is related to the action J by

$$J = 2\int_{q_-(E)}^{q_+(E)} \sin^{-1} \{ E - \sin q + w_1 q \}^{1/2} dq \tag{11}$$

where $q_-(E)$, $q_+(E)$ are the minimum and maximum amplirude of the periodic orbit with energy E. The definition of W near the points of return of the orbit is accomplished by changing the sign in the square root. The canonical change of variables is now given by

$$p = W_q(q,J,w_1)$$

$$\theta = W_J(q,J,w_1)$$

It is not hard to verify that θ is a smooth function of p, q and w_1. The function θ is of course multivalued and is a distorted polar coordinate along the orbit. It also follows that the functions p and q are periodic in θ with period one.

In the new variables the Hamiltonian takes the form:

$$\overline{H} = \epsilon \hat{H}(J) + \epsilon^2 \tilde{H}(J,\theta,t) \tag{12}$$

To compute $H(J)$ we must compute the integral defining J as a function of E. This can be done approximate (close to the equilibrium point $q = -\Pi/2 + q^1$, p close to zero). For such motions the energy E is close to $-1+\Pi/2\, w_1$ and the action J, which is the area enclosed by the orbit is close to zero.

A straightforward expansion of (4) gives the relation between E and J in the form

$$E = -1+w_1\ \Pi/2 + \frac{J}{4\alpha} - \frac{\beta}{2\alpha^3}\ J^2 + 0(\frac{w_1}{J},\ w_1^2 J,\ J^3)$$

where $\alpha = \sqrt{2} \int \operatorname{Sin}^2 \theta\, d\theta$, $\beta = \frac{\sqrt{2}}{4.3} \int (1+\operatorname{Sin}^2\theta)d\theta$

and therefore $\hat{H}(J) = {}^{J}/_{4\alpha}\quad {}^{\beta}/_{2\alpha^3}\ J^2 + 0(\frac{w^1}{J}, w_1^2 J, J^3)$

The Poincaré map for $0 \leqslant t \leqslant \frac{2\Pi}{w_0}$ associated with the solution for the Hamiltonian (12) is given by:

$$J_1 = J + e^2\ f(J,\theta)$$

$$\theta_1 = \theta + e\,\hat{H}_J(J) + e^2\ g(J,\theta)$$

This Poincaré map satisfies the assumptions for Moser's theorem [7] in fact if we let $J = r$, $\alpha(r) = e(\frac{1}{4\alpha} - \frac{\beta}{\alpha^3}\ r)$, $|\alpha''(r)| \geqslant e\frac{\beta}{\alpha^3} > 0$. We can apply the theorem choosing r to be small $a_0 < r < b_0$ and taking w sufficiently small. In fact the existence of the invariant curve is guaranteed provided that the perturbation therms satisfy an inequality of the form

$$|e^2 f(J,\theta)| + |e^2 g(J,\theta)| + |0(\frac{w_1}{J},\ w_1^2 J, J^3)| \leqslant |\alpha(b_0) - \alpha(a_0)|\delta.$$

Where δ is independent of $\gamma = |\alpha(b_0) - \alpha(a_0)|$. This can be achieved choosing ϵ and w_1 suffciently small. The invariant curve lies in the region $a_0 < J < b_0$ for all values of w_1.

The desired result of boundedness of p,q follows inmediately since initial values in the region $0 \leqslant J \leqslant a_0/2$ stay for all times in the region $0 \leqslant J \leqslant b_0$ for all values of w_1 sufficiently small. Since the transformation form p,q to θ,J is an isomorphism for all values of w_1 it folows that there is a region around $p = 0$, $q = -\Pi/2$ of initial values which produce only bounded solutions for small w_1. We have thus found a region in the Φ_1, Φ_2 and variables and therefore in the φ_1, φ_2 variables for which the solutions for φ_1, φ_2 stay bounded. This completes the proof for phase locking in this hamiltonian example.

It is clear now how the same procedure will apply also to higher order averaged systems in 2 dimensions provided the averaged system has a family of periodic orbits with action dependent period.

One could attemp to prove a similar result for a Hamiltonian system in more dimensions. The averaging procedure could be performed in the same way. If the average Hamiltonian turns out to be completely integrable it is possible to show again existence of in variant tori using the K.A.M. results. However in this case the orbits do not stay inside the tori but undergo Arnold diffusion. Numerical experiments [3] suggested that the growth of φ_i is proportional to $t^{1/2}$ (the appropriate behavior for diffusion). If this is inded the case the phase locking result is again aveilable. However we have no prood of such a result even in the simplest cases [3].

BIBLIOGRAPHY

Arnold, Small Denominators I. Mappings of the circumference onto itself, A.M.S. Translation Series. 2, 46, 581-585.

Arnold and Avez, Ergodic Problems in Classical Mechanics. W.A. Benjamin, New Yok, 1968.

B. Chirikov, A universal instability of many dimensional oscillator systems. Physics Reports 52, 263-379, 1979.

J. Hale, Ordinary Differential Equations. Wiley, New York, 1969.

F. Hoppensteadt, J.P. Kenner, Phase locking and biological clocks. Journal of Mathematical Biology, 15, 339-349, 1982.

J. Kevorkian, J. Cole, Perturbation Methods in Applied Mathematics. Springer Verlag. New York, Heidelberg, Berlin, 1981.

J.K. Moser and C.L. Siegel, Lectures on Celestial Mechanics. Springer Verlag, New York, Heidelberg, Berlin, 1971.

Contemporary Mathematics
Volume **58**, Part III, 1987

GROUPES D'HOMEOMORPHISMES DU CERCLE ET COHOMOLOGIE BORNEE

Etienne GHYS

ABSTRACT. We show that the second bounded cohomology group of a discrete group Γ can be used to study the dynamics of the actions of Γ on the circle.

1- Introduction.

Pour décrire la dynamique d'un homéomorphisme du cercle respectant l'orientation, on dispose de la notion de nombre de rotation. Il s'agit d'un élément de $\mathbf{R}/\mathbf{Z}$ qui est un invariant de semi-conjugaison. La définition usuelle d'une semi-conjugaison n'étant pas symétrique, nous allons la modifier légèrement. Nous dirons qu'une application h du cercle $\mathbf{S}^1 = \mathbf{R}/\mathbf{Z}$ dans lui-même est croissante de degré 1 si elle se relève en une application croissante $\bar{h}$ de $\mathbf{R}$ dans $\mathbf{R}$ telle que, pour tout réel x, on a : $\bar{h}(x+1) = \bar{h}(x) + 1$. Deux homéomorphismes du cercle f et g seront alors dits "semi-conjugués" s'il existe une application h croissante de degré 1 telle que $f \circ h = h \circ g$. Il s'agit d'une relation d'équivalence (voir 2-1). La propriété essentielle du nombre de rotation peut alors s'exprimer de la façon suivante : deux homéomorphismes du cercle qui respectent l'orientation sont semi-conjugués si et seulement si leurs nombres de rotation sont égaux.

Nous nous proposons ici de généraliser cette notion.

1980 Mathematics Subject Classification. 58 D 05

© 1987 American Mathematical Society
0271-4132/87 $1.00 + $.25 per page

aux groupes d'homéomorphismes. Soit Γ un groupe discret quelconque et ϕ_1, ϕ_2 deux représentations de Γ dans le groupe $\text{Homéo}^+(\mathbb{S}^1)$ des homéomorphismes du cercle qui respectent l'orientation. Nous dirons évidemment que ϕ_1 et ϕ_2 sont semi-conjuguées s'il existe une application h croissante de degré 1 telle que pour tout γ de Γ on ait $\phi_1(\gamma) o h = h o \phi_2(\gamma)$. Notre but est de construire un invariant complet de semi-conjugaison pour de telles représentations.

Un invariant existe déjà: la classe d'Euler. La donnée d'une représentation de Γ dans $\text{Homéo}^+(\mathbb{S}^1)$ permet en effet de construire, par suspension, un fibré en cercles au dessus de l'espace d'Eilenberg-MacLane $K(\Gamma,1)$. La classe d'Euler de ce fibré, élément de $H^2(\Gamma;\mathbb{Z})$, est un invariant de semi-conjugaison (voir 5-3). Il est clair, cependant, que cette classe d'Euler est insuffisante pour caractériser une représentation; lorsque $\Gamma = \mathbb{Z}$, on ne retrouve pas le nombre de rotation puisque, dans ce cas, la classe d'Euler est toujours nulle ($H^2(\mathbb{Z};\mathbb{Z}) = 0$).

L'invariant que nous construisons est un élément de la cohomologie bornée de Γ à coefficients entiers. Rappelons d'abord une définition (voir [Gr] pour cette définition et ses nombreuses motivations géométriques; voir aussi [Br-Se] et [Mit]). Désignons par A le groupe additif $\mathbb{R}$ ou $\mathbb{Z}$. On considère, pour chaque entier n, l'ensemble $C_b^n(\Gamma,A)$ formé des applications bornées de Γ^n dans A. Il est facile de vérifier que la collection des $C_b^n(\Gamma,A)$ forme un sous-complexe du complexe d'Eilenberg-MacLane de Γ (bar construction). La cohomologie de ce complexe est appelée " cohomologie bornée de Γ à coefficients dans A " et notée $H_b^*(\Gamma;A)$. Observons qu'il existe un morphisme naturel de $H_b^*(\Gamma;A)$ dans la cohomologie usuelle $H^*(\Gamma;A)$, obtenu en " oubliant " qu'un cocycle est borné.

Le fait que la classe d'Euler puisse être représentée par un cocycle borné a déjà été observé par plusieurs auteurs (voir [Mil],[Wo],[Su],[Gr],[Je], ...). Notre contribution consiste à montrer que si l'on se " rappelle " que la classe d'Euler est bornée, on obtient presque toute l'information relative à la dynamique de l'action de Γ sur $\mathbb{S}^1$.

<u>Théorème A</u> : Il existe une classe e dans $H_b^2(\text{Homéo}^+(\mathbb{S}^1);\mathbb{Z})$ ayant les propriétés suivantes:

1) Deux représentations ϕ_1 et ϕ_2 d'un groupe discret Γ dans $\text{Homéo}^+(\mathbb{S}^1)$ sont semi-conjuguées si et seulement si les deux éléments $\phi_1^*(e)$ et $\phi_2^*(e)$ de $H_b^2(\Gamma;\mathbb{Z})$ sont égaux.

2) L'image de e par le morphisme naturel de $H_b^2(\text{Homéo}^+(\mathbb{S}^1);\mathbb{Z})$ dans $H^2(\text{Homéo}^+(\mathbb{S}^1);\mathbb{Z})$ n'est autre que la classe d'Euler.

3) On a $H_b^2(\mathbb{Z};\mathbb{Z}) = \mathbb{R}/\mathbb{Z}$ et si $\phi: \mathbb{Z} \longrightarrow \text{Homéo}^+(\mathbb{S}^1)$ est une représentation, alors $\phi^*(e) \in \mathbb{R}/\mathbb{Z}$ est le nombre de rotation de l'homéomorphisme $\phi(1)$.

Puisqu'une représentation est caractérisée par un

élément de $H_b^2(\Gamma;\mathbf{Z})$, il est naturel de se demander, inversement, quels sont les éléments de $H_b^2(\Gamma;\mathbf{Z})$ qui correspondent à une représentation. Le théorème suivant répond à cette question.

Théorème B: Soit Γ un groupe dénombrable discret et z un élément de $H_b^2(\Gamma;\mathbf{Z})$. Il existe une représentation ϕ de Γ dans $\text{Homéo}^+(\mathbf{S}^1)$ telle que $\phi^*(e) = z$ si et seulement si z peut être représenté par un 2-cocycle qui ne prend que les valeurs 0 et 1.

Cet article est organisé de la façon suivante; le §2 traite de la notion de semi-conjugaison et le §3 donne deux exemples de calculs de cohomologie bornée. Le §4 est préparatoire au §5 où l'on démontre le théorème A. Enfin, le théorème B est démontré au §6.

2- Semi-conjugaisons d'homéomorphismes du cercle.

Commençons par démontrer un résultat énoncé dans l'introduction.

Proposition 2-1 : La relation de semi-conjugaison entre représentations d'un groupe dans $\text{Homéo}^+(\mathbf{S}^1)$,telle qu'elle a été définie dans l'introduction, est une relation d'équivalence.

<u>Démonstration</u> : Seule la symétrie pose problème. Supposons donc que l'on dispose de deux représentations ϕ_1 et ϕ_2 du groupe Γ dans Homéo$^+(\mathbf{S}^1)$ et d'une application h croissante de degré 1 de $\mathbf{S}^1$ dans $\mathbf{S}^1$ telle que, pour tout γ de Γ , on ait $\phi_1(\gamma) o h = h o \phi_2(\gamma)$. Soit $\bar{h}: \mathbb{R} \to \mathbb{R}$ un relevé croissant de h tel que $\bar{h}(x+1) = \bar{h}(x)+1$. Considérons alors l'application $\bar{h}^*$ de $\mathbb{R}$ dans $\mathbb{R}$ définie par:

$$\bar{h}^*(x) = \operatorname{Sup}\{ y \in \mathbb{R} \mid \bar{h}(y) \leqslant x \}$$

Il est clair que $\bar{h}^*$ est croissante et vérifie $\bar{h}^*(x+1) = \bar{h}^*(x)+1$. Par passage au quotient, $\bar{h}^*$ définit donc une application h^* du cercle dans lui même, croissante de degré 1. Par construction de h^*, et puisque $\phi_1(\gamma) o h = h o \phi_2(\gamma)$, il est facile de vérifier que $h^* o \phi_1(\gamma) = \phi_2(\gamma) o h^*$. Ceci montre que la relation de semi-conjugaison est bien une relation d'équivalence. ||

Le but des propositions qui suivent est de préciser la signification dynamique de la notion de semi-conjugaison. Rappelons que si $\phi : \Gamma \to$ Homéo$^+(\mathbf{S}^1)$ est une représentation, trois cas sont possibles:

1) ϕ possède une orbite finie. Dans ce cas, toutes les orbites finies ont le même cardinal.

2) ϕ est minimale, i.e. toutes ses orbites sont denses dans $\mathbf{S}^1$.

3) ϕ possède un " minimal exceptionnel ", i.e. il existe un ensemble de Cantor $K \subset S^1$ qui est invariant par ϕ et tel que l'orbite de tout point de K est dense dans K. Un tel minimal K, s'il existe, est nécessairement unique (voir [He-Hi] Chap IV-3).

Essentiellement, la notion de semi-conjugaison s'avère extrémement puissante dans les cas 2) et 3) et assez faible dans le cas 1) puisqu'elle ne permet alors que de déterminer le comportement des orbites périodiques.

Proposition 2-2 : Soient ϕ_1 et ϕ_2 deux représentations du groupe Γ dans $\text{Homéo}^+(S^1)$ et soit $h: S^1 \to S^1$ une application croissante de degré 1 qui est une semi-conjugaison entre ϕ_1 et ϕ_2 ,i.e. $\phi_1(\gamma)oh = ho\phi_2(\gamma)$ pour tout γ de Γ .

1) Si ϕ_1 est minimale, alors h est injective.

2) Si ϕ_2 est minimale, alors h est surjective.

De sorte que, si ϕ_1 et ϕ_2 sont minimales, alors h est en fait une véritable conjugaison topologique entre ϕ_1 et ϕ_2.

Démonstration: Ceci résulte de la structure des applications croissantes. Si $h : S^1 \to S^1$ est croissante de degré 1, on peut définir deux ouverts U et V de S^1 de la façon suivante; U est la réunion des intérieurs des intervalles maximaux sur lesquels h est constante. Pour définir V, on constate que le complémentaire de l'image de h est une réunion disjointe d'intervalles. On définit alors V comme étant la réunion des intérieurs de ces intervalles. Si h est une semi-conjugaison entre ϕ_1 et ϕ_2 , il est clair que U est invariant par ϕ_1 et que V est invariant par ϕ_2. Si ϕ_1 est minimale, U doit être vide car, h n'étant pas constante, U ne peut être le cercle tout entier. Dans ces conditions, h n'est constante sur aucun intervalle et doit donc être injective car elle est de degré 1. De même, si ϕ_2 est minimale, V doit être vide car V n'est certainement pas le cercle tout entier. Dans ce cas, h est surjective. ||

<u>Proposition 2-3</u>: Soit $\phi_1 : \Gamma \longrightarrow \text{Homéo}^+(S^1)$ une représentation possédant une orbite finie $O_1 \subset S^1$. Alors, une représentation $\phi_2 : \Gamma \longrightarrow \text{Homéo}^+(S^1)$ est semi-conjuguée à ϕ_1 si et seulement si :

1) ϕ_2 a une orbite finie O_2 de même cardinal que O_1.

2) Il existe une " bijection ordonnée " de O_1 sur O_2 qui est équivariante sous les actions de ϕ_1 et ϕ_2 .

<u>Démonstration</u>: Supposons que ϕ_2 satisfait les conditions 1) et 2). Notons ($a_1,\dots,a_n$) les éléments de O_1 ordonnés cycliquement et ($b_1,\dots,b_n$) les images de ceux ci par la bijection de O_1 sur O_2 dont on dispose. Considérons alors l'application h croissante de degré 1 qui envoie les intervalles $[a_1,a_2[,\dots,[a_n,a_1[$ sur les points $b_1,\dots,b_n$. Il est clair que h est une semi-conjugaison entre ϕ_1 et ϕ_2.

Réciproquement, soit h une semi-conjugaison entre ϕ_2 et ϕ_1 , i.e. $\phi_2(\gamma) o h = h o \phi_1(\gamma)$. L'image de O_1 par h est une orbite finie O_2 pour ϕ_2 et il nous reste à montrer que la restriction de h à O_1 est une bijection de O_1 sur O_2. Si tel n'était pas le cas, O_2 aurait un cardinal strictement inférieur à celui de O_1. Si h^* désigne maintenant une application croissante de degré 1 telle que $\phi_1(\gamma) o h^* = h^* o \phi_2(\gamma)$, $h^*(O_2)$ serait une orbite finie de ϕ_1 qui aurait strictement moins d'éléments que O_1, ce qui est impossible. ||

$$\alpha(\gamma_1,\gamma_2) = \text{Aire du triangle géodésique } (x,\gamma_1 x,\gamma_2 x) .$$

Comme tout triangle géodésique a une aire inférieure à π , il s'agit d'un cocycle borné, de norme inférieure à π. Considérons alors une triangulation de Σ par des triangles géodésiques de sommets au point de Σ correspondant à x. Si l'on interprète ces triangles comme des simplexes du complexe d'Eilenberg-MacLane de $\pi_1(\Sigma)$, on obtient:

$$\alpha([\Sigma]) = \text{Aire}(\Sigma) = -2\pi\chi(\Sigma) \quad \text{(Gauss-Bonnet)}.$$

Par conséquent, la classe de $-\frac{1}{2\pi\chi(\Sigma)}\alpha$ dans $H_b^2(\pi_1(\Sigma);\mathbb{R})$ se projette dans $H^2(\pi_1(\Sigma);\mathbb{R})$ sur c_Σ . Ceci montre que c_Σ est dans la partie bornée de la cohomologie de $\pi_1(\Sigma)$.

Puisque la norme de $-\frac{1}{2\pi\chi(\Sigma)}\alpha$ est inférieure à $-\frac{\pi}{2\pi\chi(\Sigma)}$, il nous reste à montrer que si α' est un cocycle borné se projettant sur c_Σ, alors $||\alpha'|| \geqslant -\frac{1}{2\chi(\Sigma)}$. Soit g le genre de Σ et $\Sigma' \to \Sigma$ un revêtement à k feuillets. Le genre g' de Σ' est tel que $2g'-2 = k(2g-2)$. On peut alors représenter $k[\Sigma]$ par une somme de $4g'-2$ simplexes. On a alors:

$$\alpha'(k[\Sigma]) = k \leqslant (4g'-2)||\alpha'||$$

$$||\alpha'|| \geqslant \frac{k}{4g'-2} = \frac{k}{4kg-4k+2} \xrightarrow[k\to\infty]{} -\frac{1}{2\chi(\Sigma)} .$$

Ceci démontre l'inégalité souhaitée.

La même formule montre que si $\chi(\Sigma) = 0$ (i.e. Σ est un tore), il ne peut exister de classe de cohomologie bornée α' qui se projette sur c_Σ . En effet, on a dans ce cas g=1 , de sorte que l'inégalité devient:

$$||\alpha'|| \geqslant \frac{k}{2} \xrightarrow[k\to+\infty]{} +\infty$$

ce qui est impossible. Ceci montre que si $\chi(\Sigma) = 0$, alors c_Σ n'est pas dans la partie bornée de la cohomologie de $\pi_1(\Sigma)$. ||

Remarque 3-2 : La première partie de la proposition peut être généralisée de la façon suivante: si Γ est un groupe moyennable, alors $H_b^*(\Gamma;\mathbb{R}) = 0$ (Voir [Gr]).

La borne supérieure du module d'un élément de $C_b^n(\Gamma;A)$ permet de définir, de manière naturelle, une " semi-norme" , notée $\|\ \ \|$, sur les groupes $H_b^*(\Gamma;A)$. L'image du morphisme de $H_b^*(\Gamma;A)$ dans $H^*(\Gamma;A)$ sera appelée " la partie bornée de la cohomologie de Γ à coefficients dans A ". Cette partie bornée est, elle aussi, munie d'une " semi-norme " naturelle, que nous noterons encore $\|\ \ \|$.

Notre second exemple est déjà traité dans [Gr] (dans le cadre de l'homologie singulière). Nous en donnons ici une démonstration dans notre langage.

Proposition 3-3 : Soit Σ une surface compacte orientable différente de la sphère, $\chi(\Sigma)$ sa caractéristique d'Euler-Poincaré, $[\Sigma]$ sa classe fondamentale dans $H_2(\Sigma;\mathbb{R}) \sim H_2(\pi_1(\Sigma);\mathbb{R})$ et c_Σ la classe fondamentale duale dans $H^2(\pi_1(\Sigma);\mathbb{R})$ (i.e. $c_\Sigma([\Sigma]) = 1$). Alors, c_Σ est dans la partie bornée de la cohomologie de $\pi_1(\Sigma)$ à coefficients réels si et seulement si $\chi(\Sigma)$ est non nulle. La semi-norme de c_Σ est alors $-\dfrac{1}{2\chi(\Sigma)}$.

Démonstration : Supposons d'abord que $\chi(\Sigma)$ est non nulle et munissons Σ d'une métrique à courbure -1, ce qui permet d'identifier $\pi_1(\Sigma)$ à un sous-groupe du groupe des isométries du disque de Poincaré D^2. Soit x un point de D^2. Définissons un 2-cocycle α sur $\pi_1(\Sigma)$ par :

3- Deux exemples.

Nous proposons dans ce paragraphe deux exemples de calculs de cohomologie bornée qui nous seront utiles par la suite.

Proposition 3-1 : $H_b^2(\mathbb{Z};\mathbb{R}) = 0$; $H_b^2(\mathbb{Z};\mathbb{Z}) = \mathbb{R}/\mathbb{Z}$.

Démonstration: Puisque $H^2(\mathbb{Z};\mathbb{R}) = 0$, un 2-cocycle de $\mathbb{Z}$ est une fonction $c: \mathbb{Z}^2 \to \mathbb{R}$ du type :

$$c(n,p) = du(n,p) = u(n+p) - u(n) - u(p)$$

où u est une fonction de $\mathbb{Z}$ dans $\mathbb{R}$ a priori non bornée.

Le cocycle c étant supposé borné, il existe une constante C telle que:

$$| u(n+p) - u(n) - u(p) | \leq C$$

Ceci implique l'existence d'un unique réel θ tel que :

$$v(n) = u(n) - \theta n$$

soit borné (voir [La] exercice 20 page 383). On a alors:

$$dv(n,p) = v(n+p) - v(n) - v(p) = c(n,p)$$

Autrement dit, c est nul dans $H_b^2(\mathbb{Z};\mathbb{R})$.

Etudions maintenant $H_b^2(\mathbb{Z};\mathbb{Z})$. Si θ est un réel, on considère le 2-cocycle entier c_θ de $\mathbb{Z}$ défini par:

$$c_\theta(n,p) = [\theta n+\theta p] - [\theta n] - [\theta p]$$

où [ ] désigne la partie entière. Ce cocycle est borné car il ne prend que les valeurs 0 ou 1. L'observation précédente montre que tout cocycle borné entier de $\mathbb{Z}$ est cohomologue (dans $H_b^2(\mathbb{Z};\mathbb{Z})$) à un certain c_θ . On vérifie par ailleurs que c_θ et c_ζ sont cohomologues si et seulement si $\theta-\zeta$ est un entier. ||

4- La section canonique et l'inégalité de Milnor-Wood.

Nous noterons dorénavant G le groupe Homéo$^{+}(\mathbb{S}^1)$ et $\overline{G}$ son revêtement universel, c'est-à-dire le groupe des homéomorphismes de $\mathbb{R}$ qui commutent avec les translations d'amplitudes entières. On a donc une extension centrale:

$$0 \longrightarrow \mathbb{Z} \longrightarrow \overline{G} \xrightarrow{p} G \longrightarrow 0$$

La classe d'Euler est l'obstruction à trouver un homomorphisme qui soit une section de p. Soit σ une section de p (qui n'est pas un morphisme). Considérons, pour g et h deux éléments de G, l'élément

$$c(g,h) = \sigma(fg)^{-1}\sigma(f)\sigma(g) \quad .$$

Cet élément est dans le noyau de p. C'est donc un entier. Nous obtenons donc un 2-cocycle de G dont la classe de cohomologie dans $H^2(G;\mathbb{Z})$ n'est autre que la classe d'Euler à laquelle nous nous intéressons.

Dans notre cas, il existe une façon naturelle de choisir une " section canonique " de p. Soit T le générateur du centre de $\overline{G}$ correspondant à la translation d'amplitude +1. Si g est un élément de G, les divers éléments de $p^{-1}(g)$ diffèrent des puissances de T ; il existe donc un unique élément $\overline{g}$ de $p^{-1}(g)$ tel que $\overline{g}(0) \in [0,1[$. La section canonique σ est alors celle définie par $\sigma(g) = \overline{g}$.

<u>Proposition 4-1</u> : Le cocycle c associé à ce choix de σ ne prend que les valeurs 0 et 1 .

Démonstration : Soient g et h deux éléments de G. Par définition de la section canonique, on a :

$$\overline{h}(0) \in [0,1[\ .$$

Puisque $\overline{g}$ est croissant, on a :

$$\overline{g}(\overline{h}(0)) \in [\overline{g}(0),\overline{g}(1)[\ .$$

L'intervalle $[\overline{g}(0),\overline{g}(1)[$ est d'amplitude 1 et $\overline{g}(0) \in [0,1[$. On a donc:

$$\overline{g}\overline{h}(0) \in [0,2[.$$

Puisque $\overline{g}\overline{h}$ et $\overline{gh}$ sont deux relevés du même élément de G et que $\overline{gh}(0) \in [0,1[$, on a l'égalité

$$\overline{g}\overline{h} = \overline{gh}\ T^{\varepsilon}$$

où $\varepsilon = 0$ ou 1 . Ceci signifie que c(g,h) ne peut être égal qu'à 0 ou 1. ||

Corollaire 4-2 : La classe d'Euler peut être représentée par un cocycle borné réel de norme inférieure à 1/2.

Démonstration : Remarquons que le cocycle k défini par $k(g,h) = 1/2$ est le cobord de la cochaîne constante égale à $-1/2$. La classe d'Euler est donc représentée par le cocycle c-k qui ne prend que les valeurs +1/2 et -1/2 . ||

Corollaire 4-3 : (Inégalité de Milnor-Wood) Soit ϕ une représentation du groupe fondamental d'une surface compacte orientable Σ dans le groupe des homéomorphismes du cercle respectant l'orientation. Soit $eu(\phi)$ le nombre d'Euler du fibré en cercles associé à ϕ . Alors

$$|eu(\phi)| \leqslant |\chi(\Sigma)| \ .$$

<u>Démonstration</u> : On reprend les notations de 3-3. La classe d'Euler du fibré considéré est $eu(\phi)\ c_\Sigma$. Si $\chi(\Sigma) \neq 0$, la proposition 3-3 et le corollaire précédent montrent que:

$$|eu(\phi)\ (-\frac{1}{2\chi(\Sigma)})| \leqslant \frac{1}{2} .$$

Ceci donne l'inégalité annoncée. Si $\chi(\Sigma) = 0$, nous savons que $eu(\phi)\ c_\Sigma$ est une classe de cohomologie bornée et que c_Σ ne l'est pas. Dans ce cas, $eu(\phi)$ est donc nul. ||

5- Démonstration du théorème A.

Puisque le 2-cocycle c que nous avons construit au paragraphe précédent ne prend que les valeurs 0 et 1 ,nous pouvons considérer sa classe de cohomologie dans $H_b^2(G;\mathbb{Z})$. Nous allons montrer que cette classe, notée e, vérifie les propriétés énoncées dans le théorème A.

Une partie du théorème est déjà claire:

<u>Proposition 5-1</u> : L'image de e par le morphisme naturel de $H_b^2(G;\mathbb{Z})$ dans $H^2(G;\mathbb{Z})$ est la classe d'Euler. ||

Nous abordons maintenant la démonstration de la partie essentielle du théorème A.

<u>Proposition 5-2</u> : Soient ϕ_1 et ϕ_2 deux représentations d'un groupe Γ dans G. Si ϕ_1 et ϕ_2 sont semi-conjuguées, alors les deux éléments $\phi_1^*(e)$ et $\phi_2^*(e)$ de $H_b^2(\Gamma;\mathbb{Z})$ sont égaux.

Démonstration : Soit h une semi-conjugaison entre ϕ_1 et ϕ_2 et $\bar{h}$ l'application croissante de $\mathbf{R}$ dans $\mathbf{R}$ qui relève h et telle que $\bar{h}(0) \in [0,1[$. Par hypothèse, pour tout γ de Γ, on a :

$$\phi_1(\gamma)h = h\phi_2(\gamma) .$$

Les deux applications croissantes $\overline{\phi_1(\gamma)}\bar{h}$ et $\bar{h}\overline{\phi_2(\gamma)}$ correspondent aux mêmes applications croissantes de degré 1 de S^1 dans S^1. Il existe donc une fonction u de Γ dans $\mathbf{Z}$ telle que :

$$\overline{\phi_1(\gamma)}\bar{h} = \bar{h}\overline{\phi_2(\gamma)}T^{u(\gamma)} \qquad (1)$$

Revenant à la définition du 2-cocycle c, on obtient:

$$\phi_1^*(c)(\gamma,\gamma') - \phi_2^*(c)(\gamma,\gamma') = u(\gamma\gamma') - u(\gamma) - u(\gamma') .$$

Pour démontrer que $\phi_1^*(e)$ et $\phi_2^*(e)$ sont égaux dans $H_b^2(\Gamma;\mathbf{Z})$, il suffit donc de montrer que u est bornée. Puisque

$$\bar{h}(0) \in [0,1[$$

$$\overline{\phi_1(\gamma)}(0) \in [0,1[$$

on a:

$$\overline{\phi_1(\gamma)}\bar{h}(0) \in [0,2[.$$

De même,

$$\bar{h}\overline{\phi_2(\gamma)}(0) \in [0,2[.$$

L'équation (1) montre alors que u ne peut prendre que les valeurs -1 , 0 ou +1 . ||

Remarque 5-3 : Evidemment, cette proposition montre, en particulier, que la classe d'Euler est un invariant de semi-conjugaison.

Proposition 5-4 : Réciproquement, si ϕ_1 et ϕ_2 sont deux représentations de Γ dans G telles que $\phi_1^*(e)$ et $\phi_2^*(e)$ sont égaux dans $H_b^2(\Gamma;\mathbb{Z})$, alors ϕ_1 et ϕ_2 sont semi-conjuguées.

Démonstration : On suppose qu'il existe une fonction bornée u de Γ dans $\mathbb{Z}$ telle que :

$$\phi_2^*(c)(\gamma,\gamma') - \phi_1^*(c)(\gamma,\gamma') = u(\gamma\gamma') - u(\gamma) - u(\gamma) .$$

En particulier, les 2-cocycles $\phi_1^*(c)$ et $\phi_2^*(c)$ sont cohomologues dans $H^2(\Gamma;\mathbb{Z})$. On peut donc construire une extension centrale

$$0 \longrightarrow \mathbb{Z} \longrightarrow \overline{\Gamma} \xrightarrow{\pi} \Gamma \longrightarrow 0$$

munie de deux sections s_1 et s_2 de π et ayant les deux propriétés suivantes. D'une part, si δ désigne l'image de 1 dans $\overline{\Gamma}$, on a:

$$s_2(\gamma) = s_1(\gamma)\delta^{u(\gamma)} .$$

D'autre part, les 2-cocycles de Γ associés aux sections s_1 et s_2 sont précisément $\phi_1^*(c)$ et $\phi_2^*(c)$.

Nous pouvons alors construire deux représentations $\overline{\phi}_1$ et $\overline{\phi}_2$ de $\overline{\Gamma}$ dans $\overline{G}$ en définissant:

$$\overline{\phi}_1(\delta) = T \qquad \overline{\phi}_2(\delta) = T$$

$$\overline{\phi}_1(s_1(\gamma)) = \overline{\phi_1(\gamma)} \qquad \overline{\phi}_2(s_2(\gamma)) = \overline{\phi_2(\gamma)} .$$

Avant de terminer la démonstration, nous allons démontrer le lemme suivant:

Lemme 5-5 : Pour tout x de $\mathbb{R}$, l'application

$$\alpha \in \overline{\Gamma} \longmapsto \overline{\phi}_1(\alpha)^{-1}\overline{\phi}_2(\alpha)(x) \in \mathbb{R}$$

a une image bornée.

Démonstration du lemme : Puisque les applications $\overline{\phi}_1(\alpha)$ et $\overline{\phi}_2(\alpha)$ commutent avec T, il suffit de démontrer le lemme lorsque x est un réel de $[0,1[$ De même, puisque $\overline{\phi}_1(\alpha)$ et $\overline{\phi}_2(\alpha)$ sont croissantes, il suffit de démontrer le lemme lorsque x est nul.

Le point $\overline{\phi}_1(\alpha)^{-1}\ \overline{\phi}_2(\alpha)(0)$ ne dépend évidemment que de la projection de α dans Γ car $\overline{\phi}_1(\delta) = \overline{\phi}_2(\delta) = T$ et T est dans le centre de $\overline{G}$. Par ailleurs, si $\alpha = s_2(\gamma)$,on a :

$$\overline{\phi}_1(\alpha) = \overline{\phi}_1(s_2(\gamma)) = \overline{\phi}_1(s_1(\gamma)\delta^{u(\gamma)}) = \overline{\phi_1(\gamma)}\ T^{u(\gamma)}$$

de sorte que :

$$\overline{\phi}_1(\alpha)(0) \in [u(\gamma),u(\gamma)+1[$$

car

$$\overline{\phi_1(\gamma)}(0) \in [0,1[\ .$$

Par conséquent,

$$\overline{\phi}_1(\alpha)^{-1}(0) \in [-u(\gamma)-1,-u(\gamma)[\quad .$$

Puisque, par ailleurs:

$$\overline{\phi}_2(\alpha)(0) = \overline{\phi_2(\gamma)}(0) \in [0,1[\quad ,$$

on a :

$$\overline{\phi}_1(\alpha)^{-1}\ \overline{\phi}_2(\alpha)(0) \in [-u(\gamma)-1,-u(\gamma)+1[\quad .$$

Ceci démontre le lemme puisque, par hypothèse, u est bornée. ||

Fin de la démonstration de la proposition 5-4 :

On définit une fonction $\overline{h}$ de $\mathbb{R}$ dans $\mathbb{R}$ par :

$$\overline{h}(x) = \sup_{\alpha\in\overline{\Gamma}}(\overline{\phi}_1(\alpha)^{-1}\ \overline{\phi}_2(\alpha)(x)) \quad .$$

Il est clair que $\overline{h}$ est une fonction croissante qui commute avec T. Si α est un élément de Γ , on a :

$$\begin{aligned}\overline{h}(\overline{\Phi}(\alpha)(x) &= \sup_{\beta\in\Gamma}(\overline{\Phi}(\beta)^{-1}\ \overline{\Phi}(\beta\alpha)(x))\\ &= \sup_{\beta\in\Gamma}(\overline{\Phi}(\beta\alpha^{-1})^{-1}\ \overline{\Phi}(\beta)(x))\\ &= \overline{\Phi}_1(\alpha)(\sup_{\beta\in\Gamma}(\overline{\Phi}_1(\beta)^{-1}\ \overline{\Phi}_2(\beta)(x))\\ &= \overline{\Phi}_1(\alpha)(\overline{h}(x)) \quad .\end{aligned}$$

C'est-à-dire que l'on a

$$\overline{h}\ \overline{\Phi}_2 = \overline{\Phi}_1\ \overline{h} \quad .$$

Si l'on note h l'application croissante de degré 1 de $\mathbf{S}^1$ dans $\mathbf{S}^1$ qui est obtenue à partir de $\overline{h}$, on a :

$$h\ \phi_2 = \phi_1\ h \quad .$$

En d'autres termes, les représentations ϕ_1 et ϕ_2 sont semi-conjuguées. Ceci termine la démonstration de 5-4. ||

Pour terminer la démonstration du théorème A, il nous reste à démontrer la :

<u>Proposition 5-6</u> : Si $\Gamma = \mathbf{Z}$ et si $\phi: \Gamma \longrightarrow \mathbf{Z}$ est une représentation, alors l'élément $\phi^*(e)$ de $H_b^2(\Gamma;\mathbf{Z}) \in \mathbf{R}/\mathbf{Z}$ est le nombre de rotation de $\phi(1)$.

<u>Démonstration</u> : Soit $\theta \in \mathbf{R}/\mathbf{Z}$ et R_θ la rotation " d'angle " θ définie par $R_\theta(x) = x + \theta \pmod{\mathbf{Z}}$. Soit $\phi_\theta: \mathbf{Z} \longrightarrow G$ la représentation envoyant 1 sur R_θ . On vérifie immédiatement que le 2-cocycle $\phi_\theta^*(c)$ est égal au 2-cocycle c_θ défini en 3-1. On obtient la proposition en remarquant que toute représentation

$\phi:\Gamma \longrightarrow \mathbf{Z}$ est semi-conjuguée à la représentation ϕ_θ où θ est le nombre de rotation de $\phi(1)$. ||

Remarque 5-7: Il est clair que, pour obtenir le théorème A, il nous faut considérer la cohomologie bornée à coefficients entiers et non pas à coefficients réels. Nous avons vu, en effet, que $H_b^2(\mathbf{Z};\mathbf{R}) = 0$ de sorte que la cohomologie bornée réelle ne peut " contenir " la notion de nombre de rotation. Dans certains cas, cependant, la cohomologie bornée de Γ à coefficients réels peut suffire pour décrire les représentations de Γ dans G. Ce sera le cas si Γ est un groupe parfait car, dans ce cas, on vérifie que $H_b^2(\Gamma;\mathbf{Z})$ s'injecte dans $H_b^2(\Gamma;\mathbf{R})$.

6- Démonstration du théorème B .

Une partie du théorème B est déjà claire : si $\phi:\Gamma \longrightarrow G$ est une représentation, la classe $\phi^*(e)$ est représentée par le 2-cocycle $\phi^*(c)$ qui ne prend que les valeurs 0 et 1 d'après la proposition 4-1.

Réciproquement, soit ω un 2-cocycle sur Γ qui ne prend que les valeurs 0 et 1. On peut alors construire une extension centrale

$$0 \longrightarrow \mathbf{Z} \longrightarrow \overline{\Gamma} \underset{s}{\overset{\pi}{\rightleftarrows}} \Gamma \longrightarrow 0$$

et une section s telles que le cocycle ω soit défini par:

$$\omega(\gamma_1,\gamma_2) = s(\gamma_1\gamma_2)^{-1}s(\gamma_1)s(\gamma_2) \quad .$$

On note encore δ l'image de +1 dans $\bar{\Gamma}$. La donnée de s permet d'identifier $\bar{\Gamma}$, comme ensemble, à $\Gamma\times\mathbb{Z}$; la loi de multiplication dans ces coordonnées étant:

$$(\gamma_1,n_1)(\gamma_2,n_2) = (\gamma_1\gamma_2,n_1+n_2+\omega(\gamma_1,\gamma_2)) \qquad (2)$$

Nous ferons tout d'abord l'hypothèse que ω est un cocycle non dégénéré, c'est-à-dire qu'il satisfait la condition:

$$\omega(\gamma,1) = \omega(1,\gamma) = 0$$

(Le symbole 1 désignera aussi bien l'élément neutre de Γ que celui de $\bar{\Gamma}$, ainsi que, bien entendu, l'entier 1 ...)
Cette dernière condition est équivalente au fait que s(1) = 1 ou encore que (1,0) est l'élément neutre de $\bar{\Gamma}$. Nous nous débarasserons de cette hypothèse de non dégénérescence à la fin de ce paragraphe.

<u>Lemme 6-1</u> : Il existe une relation de préordre sur $\bar{\Gamma}$,notée $\preccurlyeq$ telle que:

i) $(1,0) \preccurlyeq (\gamma,n)$ si et seulement si $0 \leq n$;

ii) si $\alpha_1 \preccurlyeq \alpha_2$, alors pour tout α de $\bar{\Gamma}$, on a $\alpha\alpha_1 \preccurlyeq \alpha\alpha_2$;

iii) $\preccurlyeq$ est une relation de préordre total.

<u>Démonstration</u> : Une relation de préordre satisfaisant ii) est parfaitement définie par l'ensemble des éléments supérieurs à l'élément neutre qui doit être un semi-groupe. Puisque ω ne prend que des valeurs positives, la relation (2) montre

que l'ensemble des éléments de $\overline{\Gamma}$ du type (γ,n) avec $n \geqslant 0$ est effectivement un semi-groupe. Les conditions i) et ii) définissent donc bien un préordre sur $\overline{\Gamma}$.

Puisque ω ne prend que les valeurs 0 et 1, l'inverse de l'élément (γ,n) de $\overline{\Gamma}$ est soit $(\gamma^{-1},-n)$ soit $(\gamma^{-1},-n-1)$. En observant que, pour tout entier n, on a:

$$0 \leqslant n \quad \text{ou} \quad (\ 0 \leqslant -n \quad \text{et} \quad 0 \leqslant -n-1 \ ) \ ,$$

on déduit que, pour tout élément (γ,n) de $\overline{\Gamma}$, on a :

$$(1,0) \preccurlyeq (\gamma,n) \quad \text{ou} \quad (1,0) \preccurlyeq (\gamma,n)^{-1}$$

ou encore, d'après ii) :

$$(1,0) \preccurlyeq (\gamma,n) \quad \text{ou} \quad (\gamma,n) \preccurlyeq (1,0) \quad .$$

Ceci montre que $\preccurlyeq$ est une relation de préordre total. ||

<u>Lemme 6-2</u> : On suppose Γ dénombrable. Alors, il existe une application i de $\overline{\Gamma}$ dans $\mathbb{R}$ telle que :

i) $i(1) = 0$;

ii) pour tout α de $\overline{\Gamma}$, on a $i(\delta\alpha) = i(\alpha) + 1$;

iii) $\alpha_1 \preccurlyeq \alpha_2$ si et seulement si $i(\alpha_1) \leqslant i(\alpha_2)$.

<u>Démonstration</u> : Soit $(1 , \gamma_1 , \gamma_2 , \dots , \gamma_n , \dots)$ une énumération de Γ. On suppose donnée une application i de $\pi^{-1}\{1,\gamma_1,\gamma_2,\dots,\gamma_n\}$ dans $\mathbb{R}$ satisfaisant les conditions i), ii) et iii). L'ensemble $\pi^{-1}\{\gamma_{n+1}\}$ est constitué de $s(\gamma_{n+1})$ et de ses translatés par les puissances de δ. Puisque la relation de préordre $\preccurlyeq$ est totale, il est facile de choisir un réel, noté $i(s(\gamma_{n+1}))$, de façon à satisfaire la condition iii) sur l'ensemble $\pi^{-1}\{1,\gamma_1,\dots,\gamma_n\} \cup s(\gamma_{n+1})$. On définit alors,

pour chaque entier k,

$$i(s(\gamma_{n+1})\,\delta^k) = i(s(\gamma_{n+1})) + k$$

de sorte que i est maintenant définie sur $\pi^{-1}\{1,\gamma_1,\ldots,\gamma_{n+1}\}$ et vérifie toujours i), ii) et iii). Par récurrence, on construit donc l'application cherchée. ||

Remarquons que $\overline{\Gamma}$ opère sur lui-même par translations à gauche et que ces translations sont des bijections croissantes de l'ensemble préordonné $(\overline{\Gamma}, \preccurlyeq)$. Soit $\overline{F}$ l'adhérence de $i(\overline{\Gamma})$ dans $\mathbb{R}$ et α un élément de $\overline{\Gamma}$. On définit une fonction $\overline{\phi}(\alpha)$ de $\overline{F}$ dans $\overline{F}$ par:

$$\overline{\phi}(\alpha)(x) = \operatorname{Sup}\{\, i(\alpha\beta) \mid i(\beta) \leqslant x \,\} .$$

La restriction de $\overline{\phi}(\alpha)$ à $i(\overline{\Gamma})$ est une bijection car $\overline{\phi}(\alpha)(i(\beta)) = i(\alpha\beta)$. Il se peut cependant que $\overline{\phi}(\alpha)$ ne soit pas un homéomorphisme de $\overline{F}$. Le lemme suivant donne une condition pour que $\overline{\phi}(\alpha)$ soit un homéomorphisme.

<u>Lemme 6-3</u> : On suppose que les deux extrémités x et y de toute composante connexe $]x,y[$ de $\mathbb{R}-\overline{F}$ sont dans $i(\overline{\Gamma})$. Alors, pour tout α de $\overline{\Gamma}$, la fonction $\overline{\phi}(\alpha)$ est un homéomorphisme de $\overline{F}$.

<u>Démonstration</u> : Il suffit de montrer que $\overline{\phi}(\alpha)$ est continue sur $\overline{F}$ car l'inverse de $\overline{\phi}(\alpha)$ sera évidemment $\overline{\phi}(\alpha^{-1})$.

Soit t un élément de $\overline{F}$. Trois cas sont a priori possibles: t peut être un point isolé de $\overline{F}$, un point d'accumulation "unilatéral" de $\overline{F}$ ou un point d'accumulation "bilatéral" de $\overline{F}$. La question de la continuité de $\overline{\phi}(\alpha)$ en t ne se pose évidemment que dans les deux derniers cas.

Supposons donc que t est un point d'accumulation unilatéral de $\overline{F}$, par exemple à gauche. Dans ce cas, t est l'extrémité gauche d'une composante connexe de $\mathbb{R}-\overline{F}$ et, d'après l'hypothèse faite, t est du type $i(\beta_1)$ pour un certain élément β_1 de Γ. Il suffit évidemment de montrer la continuité de $\overline{\phi}(\alpha)$ à gauche. Dans le cas contraire, on aurait:

$$\lim_{\substack{t'\to t\\ t'<t}} \overline{\phi}(\alpha)(t') < \overline{\phi}(\alpha)(t)$$

Le membre de droite de cette inégalité est égal à $v = i(\alpha\beta_1)$. Quant au membre de gauche, d'après la définition de $\overline{\phi}(\alpha)$, il est égal à :

$$u = \operatorname{Sup}\{\, i(\alpha\tau) \mid \tau \prec \beta_1 \}$$

où la notation $\tau \prec \beta_1$ signifie $i(\tau) < i(\beta_1)$. L'intervalle $]u,v[$ ne peut contenir d'élément de $i(\Gamma)$. En effet, si $i(\beta_2)$ est un élément de $]u,v[$, alors l'élément $\beta_3 = \alpha^{-1}\beta_2$ aurait la propriété que $\beta_3 \prec \beta_1$ ce qui entraînerait la contradiction suivante:

$$u = \operatorname{Sup}\{\, i(\alpha\tau) \mid \tau \prec \beta_1 \} \geq i(\alpha\beta_3) = i(\beta_2) > u\ .$$

Comme u et v sont clairement des éléments de $\overline{F}$, l'intervalle $]u, v[$ est donc une composante connexe de $\mathbb{R}-\overline{F}$. D'après l'hypothèse faite, u est du type $i(\beta_4)$ pour un certain β_4 de Γ. La définition de u montre alors que β_4 est tel que $\tau \prec \beta_1$ si et seulement si $\alpha\tau \preceq \beta_4$. En d'autres termes, il n'y aurait aucun élément de $i(\Gamma)$ entre $i(\alpha^{-1}\beta_4)$ et $t = i(\beta_1)$ ce qui contredit le fait que t est un point d'accumulation à gauche de $i(\Gamma)$.

Dans le cas où t est un point d'accumulation bilatéral

de $i(\overline{T})$, on procède de façon similaire pour montrer que $\overline{\Phi}(\alpha)$ est continu à gauche et à droite. ||

<u>Lemme 6-4</u> : On suppose toujours que Γ est dénombrable. Alors, il existe une application i de $\overline{T}$ dans $\mathbb{R}$ vérifiant les conditions i), ii) et iii) du lemme 6-2 ainsi que la condition du lemme 6-3.

<u>Démonstration</u> : Soit $i_0 : \overline{T} \to \mathbb{R}$ une application vérifiant les conditions i), ii) et iii) du lemme 6-2. Nous allons définir une autre application $i : \overline{T} \to \mathbb{R}$ de la forme $i = l \circ i_0$ où $l : \mathbb{R} \to \mathbb{R}$ sera une application croissante qui "écrase sur un point" les composantes connexes de $\mathbb{R}-\overline{F}$ ne satisfaisant pas à la condition du lemme 6-3. Plus précisément, soit $\mathcal{C}$ la collection des composantes connexes de $\mathbb{R}-\overline{F}$ dont au moins une des extrémités n'est pas dans $i(\overline{T})$. Si $]u,v[$ et $]v,w[$ sont deux éléments de $\mathcal{C}$ ayant une extrémité en commun, le point v est un point isolé de $\overline{F}$ et donc un élément de $i(\overline{T})$. Dans ce cas, u et w ne sont pas des éléments de $i(\overline{T})$. Considérons alors la réunion des intervalles du type $]u,w[$ ainsi décrits et des éléments de $\mathcal{C}$ qui n'ont pas d'extrémité commune avec un autre élément de $\mathcal{C}$. On obtient ainsi un ouvert dont le complémentaire $\overline{K}$ n'a pas de point isolé. Evidemment, $\overline{K}$ est invariant par translations entières. On peut donc trouver une mesure positive μ sur $\mathbb{R}$, sans atome, dont le support est exactement $\overline{K}$, invariante par translations entières et telle que $\mu[0,1[= 1$. On considère alors la fonction $l : \mathbb{R} \to \mathbb{R}$ définie par $l(x) = \mu[0,x]$ si $x \geqslant 0$ et $l(x) = -\mu[x,0]$ si $x \leqslant 0$. Par construction même

de l, les propriétés suivantes de l sont évidentes:

1- l est continue

2- la restriction de l à $i(\mathcal{T})$ est strictement croissante

3- la restriction de l à un élément de $\mathcal{C}$ est constante

Par conséquent, l'application $i = l \circ i_0$ vérifie toutes les conditions requises. ||

Nous sommes maintenant en mesure de démontrer la partie principale du théorème B.

<u>Proposition 6-5</u> : Soit Γ un groupe dénombrable et ω un 2-cocycle non dégénéré sur Γ qui ne prend que les valeurs 0 et 1. Alors, il existe une représentation $\phi: \Gamma \to G$ telle que $\phi^*(c) = \omega$.

<u>Démonstration</u> : Soit $i : \mathcal{T} \to \mathbb{R}$ une application donnée par le lemme 6-4. Dans ce cas, d'après 6-3, $\overline{\Phi}(\alpha)$ est un homéomorphisme strictement croissant de $\overline{F}$. Prolongeons $\overline{\Phi}(\alpha)$ en un homéomorphisme de $\mathbb{R}$, encore noté $\overline{\Phi}(\alpha)$, en lui imposant d'être affine sur chaque composante connexe de $\mathbb{R}-\overline{F}$. On obtient ainsi une représentation $\overline{\Phi} : \mathcal{T} \to \overline{G}$. Par passage au quotient, on obtient finalement une représentation $\phi:\Gamma \to G$. La construction même de ϕ montre que $\phi^*(c) = \omega$. ||

<u>Fin de la démonstration du théorème B :</u> Il nous reste à nous débarasser de l'hypothèse de non dégénérescence faite sur ω .

La condition exprimant que ω est un cocycle s'écrit:

$$\omega(\gamma_1,\gamma_2) + \omega(\gamma_1\gamma_2,\gamma_3) = \omega(\gamma_2,\gamma_3) + \omega(\gamma_1,\gamma_2\gamma_3)$$

On en déduit qu'il existe un entier ν tel que, pour tout γ de Γ, on ait :

$$\omega(1,\gamma) = \omega(\gamma,1) = \nu$$

Cet entier ne peut être égal qu'à 0 ou 1 et ω est non dégénéré si et seulement si ν est nul. Supposons donc que $\nu = 1$ et définissons un 2-cocycle ω' par :

$$\omega'(\gamma_1, \gamma_2) = 1 - \omega(\gamma_1, \gamma_2) .$$

Il est clair que ω' est un 2- cocycle non dégénéré qui ne prend que les valeurs 0 et 1. Par ailleurs ω' et $-\omega$ sont cohomologues dans $H_b^2(\Gamma;\mathbf{Z})$ (car la fonction constante 1 de $C_b^2(\Gamma;\mathbf{Z})$ est le cobord de la fonction constante -1 de $C_b^1(\Gamma;\mathbf{Z})$).

D'après la proposition 6-5, il existe une représentation $\phi' : \Gamma \to G$ telle que $\phi'^*(c) = \omega'$. Si $\phi: \Gamma \to G$ désigne maintenant une représentation obtenue en conjugant ϕ' par une symétrie de $\mathbf{S}^1$ qui renverse l'orientation, $\phi^*(c)$ est évidemment cohomologue à $-\phi'^*(c)$, c'est-à-dire à $-\omega'$ et donc à ω .

Ceci montre le théorème B pour un cocycle dégénéré et termine donc la démonstration de ce théorème. ||

BIBLIOGRAPHIE

[Br-Se] R. Brooks and C. Series : Bounded cohomology of surface groups, Topology 23(1984),29-36.

[Gr] M. Gromov : Volume and bounded cohomology, Pub.Math. I.H.E.S. 56(1982),5-100.

[He-Hi] G. Hector and U. Hirsch : Introduction to the geometry of foliations, Part B , Aspects Math. (1983)

[Je] S. Jekel : Simplicial categories and foliations, preprint Boston.

[La] S. Lang : Algebra, Addison Wesley (1965)

[Mil] J. Milnor : On the existence of a connection with curvature zero, Comment.Math.Helvetici32(1958),215-223.

[Mit] Y. Mitsumatsu : Bounded cohomology and l^1 homology of surfaces, to appear in Topology.

[Su] D. Sullivan : A generalization of Milnor's inequality concerning affine foliations and affine manifolds, Comment.Math.Helvetici51(1976),183-189.

[Wo] J. Wood : Bundles with totally disconnected structure group, Comment.Math.Helv.46(1971),257-273.

Université des Sciences et Techniques de Lille I
U.E.R. de Mathématiques Pures et Appliquées
E.R.A. au C.N.R.S. 07590
59655 Villeneuve d'Ascq Cedex
FRANCE

Contemporary Mathematics
Volume 58, Part III, 1987

HARMONIC MAPS OF RIEMANN SURFACES TO QUATERNIONIC PROJECTIVE SPACES AND GRASSMANNIANS

James F. Glazebrook

ABSTRACT. In this paper we discuss classes of harmonic maps to quaternionic Grassmannians and present some examples. We also discuss the notion of J_2- holomorphicity and the twistor degree for maps to quaternionic projective space. Finally, we append with some topology of the space $(\mathbb{C}P^{2n+1}, J_2)$.

1. ISOTROPIC HARMONIC MAPS TO COMPLEX GRASSMANNIANS. Throughout this paper, M will denote a Riemann surface (open or closed). Let N be a Kähler manifold and let $\phi : M \to N$ be a C^∞ map. We consider the complexified differential $d^{\mathbb{C}}\phi : T^{\mathbb{C}}M \to T^{\mathbb{C}}N$, and with respect to complex tangent vectors of type $(1, 0)$ and $(0, 1)$ on M, denoted by $T'M$ and $T''M$ respectively, and likewise for $T^{\mathbb{C}}N$, we have a decomposition of $d^{\mathbb{C}}\phi$ into 4 globally defined maps:

$$(1.1)\qquad \begin{cases} \partial\phi : T'M \to T'N & \overline{\partial}\phi : T''M \to T'N \\ \partial\overline{\phi} : T'M \to T''N & \overline{\partial}\overline{\phi} : T''M \to T''N \end{cases}$$

We shall denote smooth sections of a bundle E by $C(E)$, and with respect to a chart $U = (U, z)$ of M, where U is a non-empty open set in M and $z: U \to \mathbb{C}$ is a complex co-ordinate, we shall write smooth sections over U as C_U. Then the above maps $\partial\phi$ and $\overline{\partial}\phi$ can be represented locally by the maps

$$(1.2)\qquad \partial'\phi = \partial\phi(\partial/\partial z) \qquad\qquad \partial''\phi = \overline{\partial}\phi(\partial/\partial\overline{z})$$

respectively with $\partial'\phi, \partial''\phi \in C_U(\phi^{-1}T'N)$. Now let ∇ denote the connection on the bundle $\phi^{-1}T'N$; we have a splitting of ∇ into types $(1, 0)$ and $(0, 1)$, namely, $\nabla' = \nabla_{\partial/\partial z}$ and $\nabla'' = \nabla_{\partial/\partial\overline{z}}$.

1980 Mathematics Subject Classification 58E20

© 1987 American Mathematical Society
0271-4132/87 $1.00 + $.25 per page

We say that ϕ is <u>harmonic</u> if and only if

(1.3) $\nabla''\partial'\phi = 0$ (or equivalently, $\nabla'\partial''\phi = 0$).

The way we have defined harmonicity is somewhat special since we have taken advantage of the domain being 1-complex dimensional; for the more generalised notion of harmonicity see [5].

We now consider the case $N = G_m(\mathbb{C}^{m+p})$, the Grassmannian of complex m-planes in $\mathbb{C}^{m+p}$. If Λ is taken to denote the universal m-plane bundle on $G_m(\mathbb{C}^{m+p})$, then it is well known that $T'G_m(\mathbb{C}^{m+p}) \cong \mathrm{Hom}(\Lambda, \Lambda^{\perp})$.

Given a C^{∞} map ϕ as above, we may form covariant derivatives of ϕ as follows:

$\nabla'^{\alpha}\phi$, $\nabla''^{\alpha}\phi \in C_U(\phi^{-1}T'G_m(\mathbb{C}^{m+p})) = C_U(\phi^{-1}\mathrm{Hom}(\Lambda, \Lambda^{\perp}))$ $\alpha = 1, 2, \ldots,$ where $\nabla'\phi = \partial'\phi$ and $\nabla''\phi = \partial''\phi$, $\nabla'^{\alpha} = \nabla'(\nabla'^{\alpha-1}\phi)$, etc.

For each $x \in M$ and $\alpha = 1, 2, \ldots,$ the values $(\nabla'^{\alpha}\phi)_x$ and $(\nabla''^{\alpha}\phi)_x$ may be regarded as linear maps $\Lambda_{\phi(x)} \to \Lambda^{\perp}_{\phi(x)} \subset \mathbb{C}^{m+p}$. In the terminology of [9], we say that ϕ is <u>isotropic</u> if the images of the linear maps $(\nabla'^{\alpha}\phi)_x$ and $(\nabla''^{\beta}\phi)_x$ are orthogonal with respect to the metric on $\phi^{-1}T'G_m(\mathbb{C}^{m+p})$, for $x \in M$, $\alpha, \beta = 1, 2, \ldots.$ We also say that ϕ is <u>full</u> if the only subspace of $\mathbb{C}^{m+p}$ containing $\mathrm{Im}\phi$, is $\mathbb{C}^{m+p}$ itself, where $\mathrm{Im}\,\phi = \bigcup_{x\in U}\left(\{x\}\times \mathrm{Im}(\phi_x)\right)$ [9].

Let V and X be holomorphic subbundles of the trivial (m+p)-plane bundle $\underline{\mathbb{C}}^{m+p}$ on M, such that

$$(1.4)\quad \begin{cases} \text{a)} & V \subset X \\ \text{b)} & \mathrm{Rank}\, X - \mathrm{Rank}\, V = m \\ \text{c)} & \partial' C(V) \subset C(X) \end{cases}$$

(Here we take ∂ to denote the trivial connection on $\mathbb{C}^{m+p}$ and $\partial' = \partial_Z$ where $Z = \partial/\partial_z$).

Now such a pair (V, X) satisfying (1.4) will be called a ∂'-pair (V, X) of rank difference m. In what follows, we assume that the ∂'-pair (V, X) is 'full' in the sense of [9].

THEOREM 1.1 [9]. *There exists a bijective correspondence between full ∂'-pairs (V, X) of $\mathbb{C}^{m+p}$ of rank difference m on M, and full, isotropic harmonic maps $\phi : M \to G_m(\mathbb{C}^{m+p})$ given by*

$$\phi(x) = V_x^{\perp} \cap X_x \qquad \text{for } x \in M . \tag{1.5}$$

Conversely, given such a ϕ, the pair $(\phi''_{(\infty)}, \tilde{\phi}''_{(\infty)})$ is a full ∂'-pair yielding (V, X) where $\phi''_{(\infty)}$ denotes a certain limiting osculating bundle of ϕ with respect to the $(0, 1)$ derivatives of ϕ and $\tilde{\phi}''_{(\infty)} = \phi + \phi''_{(\infty)}$.

Taking $m = 1$, the above theorem recovers the celebrated theorem of Eells and Wood [6] for isotropic harmonic maps $M \to \mathbb{C}P^p$. The condition $\overline{X} = V^{\perp}$ when imposed, recovers analogous maps to the real Grassmanian $G_m(\mathbb{R}^{m+p})$.

2. MAPS TO QUATERNIONIC GRASSMANIANS.

Let $\mathbb{H}$ denote the division ring of quaternions and consider the identification $\mathbb{H} \cong \mathbb{C} + \mathbb{C}j \cong \mathbb{C}^2$ where j denotes a unit quaternion. This generalises to the identification $\mathbb{H}^{n+k} \cong \mathbb{C}^{2n+2k}$, and left multiplication by j induces a conjugate linear map

$$\sigma : \mathbb{C}^{2n+2k} \to \mathbb{C}^{2n+2k} . \tag{2.1}$$

The Grassmannian $G_k(\mathbb{H}^{n+k})$ of quaternionic k-planes in $\mathbb{H}^{n+k}$

may be regarded as those complex $2k$-planes in $\mathbb{C}^{2n+2k}$ that are fixed under the action of σ, and the subsequent embedding of $G_k(\mathbb{H}^{n+k})$ into $G_{2k}(\mathbb{C}^{2n+2k})$ may be seen to be a totally geodesic embedding. In terms of the usual Hermitian inner product $\langle , \rangle$ on $\mathbb{C}^{2n+2k}$, we can define a non-degenerate skew symmetric inner product S for vectors in $\mathbb{C}^{2n+2k}$ by

$$S(x, y) = \langle x, \sigma y\rangle \qquad \text{for } x, y \in \mathbb{C}^{2n+2k}. \tag{2.2}$$

Now let us set $m = 2k$ and $p = 2n$. With regards to (1.5) the condition

$$\sigma X = V^{\perp} \tag{2.3}$$

restricts the mapping in Theorem 1.1 to have its image in $G_k(\mathbb{H}^{n+k})$. Condition (2.3) together with (1.4) readily reduces to the following conditions to be satisfied by V alone (see [10]):

$$\begin{cases} \text{a) } V \text{ as a holomorphic subbundle of } \mathbb{C}^{2n+2k} \text{ has rank } n \\ \text{b) } V \text{ is isotropic for } S \\ \text{c) } \sigma\partial' C(V) \subset C(V^{\perp}). \end{cases} \tag{2.4}$$

Regarding V as (bijectively) corresponding to a holomorphic map $\psi : M \to G_n(\mathbb{C}^{2n+2k})$, the above conditions b) and c) can be (pointwise) expressed as $S(\psi, \psi) = 0$ and $S(\psi, \partial'\psi) = 0$ respectively. Using the techniques of [9] the author proved the following theorem:

THEOREM 2.1 [10] [11]: <u>There exists a bijective correspondence between full, isotropic harmonic maps</u> $\phi : M \to G_k(\mathbb{H}^{n+k})$ <u>and full, holomorphic subbundles</u> V of $\mathbb{C}^{2n+2k}$ <u>satisfying the conditions in</u> (2.4).

The 'fullness' condition on V is a technical requirement for the proof (see [10]); the 'fullness' and 'isotropy' of ϕ is that of the harmonic map into $G_{2k}(\mathbb{C}^{2n+2k})$ on composing with the totally geodesic inclusion map. The case $k = 1$, i.e. $\mathbb{H}P^n$, was discussed in [10] and the proof there easily generalises to the Grassmannian, however, it is worth mentioning the basic idea. The construction of the maps in Theorem 2.1 concerns the homogeneous fibration

$$(2.5) \quad \pi : R_{k,n} \cong Sp(n+k)/Sp(k) \times U(n) \to G_k(\mathbb{H}^{n+k}) \cong Sp(n+k)/Sp(k) \times Sp(n).$$

The space $R_{k,n}$ is the totality of n-dimensional, S-isotropic subspaces of $\mathbb{C}^{2n+2k}$ [10] [11]. The holomorphic subbundle V in the above theorem is seen to define a harmonic map from M to $R_{k,n}$ that is horizontal with respect to π. On choosing the correct metrics, the fibration in (2.5) can be exhibited as a Riemannian submersion, and thus $\phi = \pi o \psi$ is harmonic in accordance with a fundamental composition principle [5]. Note that $R_{n,1} \cong \mathbb{C}P^{2n+1}$ which is the canonical twistor space for $\mathbb{H}P^n$, but in [10] the space $R_{1,n}$ was found to be the most suitable for the isotropic maps to $\mathbb{H}P^n$. More generally, on interchanging the roles of k and n in (2.5) we obtain the other twistor space $S_{k,n} \cong Sp(n+k)/Sp(n) \times U(k)$. These two spaces $R_{k,n}$ and $S_{k,n}$ (listed in [4] and [14]) are the two possible twistor spaces for $G_k(\mathbb{H}^{n+k})$. In fact, they each apply to two different kinds of maps classified by the Theorem 2.1: Firstly in the case of $R_{k,n}$ the proof of the theorem tells us that in the inverse assignment, the osculating bundle $\phi''_{(\infty)}$ has rank n, whereas in the case of $S_{k,n}$ it has rank k.

Remark

Quite recently, Aithal has made an independent study of harmonic maps

$S^2 \to \mathbb{H}P^2$ [2], and from the viewpoint of moving frames, Zandi has studied minimal surfaces in a general quaternionic Kähler manifold [16].

3. J_2-HOLOMORPHICITY AND THE TWISTOR DEGREE.

The twistor fibration

$$q : \mathbb{C}P^{2n+1} \to \mathbb{H}P^n \tag{3.1}$$

$$Sp(n+1)/U(1)\times Sp(n) \to Sp(n+1)/Sp(1)\times Sp(n)$$

with fibre $\mathbb{C}P^1 = Sp(1)/U(1)$, nevertheless plays an important role in the construction and classification of certain harmonic maps $\phi : M \to \mathbb{H}P^n$.

We can endow $\mathbb{C}P^{2n+1}$ with two almost complex structures as follows:
Let J^h and J^v denote the usual almost complex structures on the horizontal and vertical distributions (with respect to q) on $T\mathbb{C}P^{2n+1}$; we may define two almost complex structures J_1 and J_2 on $\mathbb{C}P^{2n+1}$ by

$$J_1 = \begin{cases} J^h \\ J^v \end{cases} \quad \text{and} \quad J_2 = \begin{cases} J^h \\ -J^v \end{cases} \tag{3.2}$$

The almost complex structure J_1 can be regarded as the 'usual' (integrable) one on $\mathbb{C}P^{2n+1}$, whereas J_2 is, in fact, never integrable [7].
Consider now a C^∞ map $\phi : M \to (\mathbb{H}P^n, h)$ where h denotes the Riemannian metric on $\mathbb{H}P^n$; letting $\partial\phi = \phi_*(\partial/\partial z)$, we say that ϕ is <u>conformal</u> if $h(\partial\phi, \partial\phi) = 0$. Further, let us say that ϕ is <u>inclusive</u> if, for all $x \in M$, $\phi_*(T_xM)$ is contained in a

quaternionic 1-dimensional subspace of the range. Then we have:

THEOREM 3.1 [7] [14]. <u>There exists a bijective correspondence between</u>

(I) <u>non-constant, inclusive conformal harmonic maps</u> $\phi : M \to \mathbb{H}P^n$, <u>and</u>

(II) <u>non-vertical</u> (i.e. <u>not contained in a fibre of</u> q) J_2-<u>holomorphic curves</u> $\psi : M \to (\mathbb{C}P^{2n+1}, J_2)$.

In fact the above theorem is true in the more general case when the range is any quaternionic Kähler manifold [15] with its canonical twistor space Z likewise endowed with a J_2 almost complex structure.

Now let us suppose that M is compact with genus g; we denote this by M_g. Also, to simplify the writing, let us set $N = \mathbb{H}P^n$ and $Z = \mathbb{C}P^{2n+1}$. For such maps ϕ, let $G \subset \phi^{-1}T^{\mathbb{C}}N$ denote the holomorphic line bundle spanned generically by $\partial\phi$ (see [13]). The globally defined form $\lambda = dz \otimes \partial\phi$ is a complex analytic section of the bundle $K_M \otimes (\phi^{-1}T^{\mathbb{C}}N)$ where K_M is the canonical line bundle of M. We define <u>the ramification index</u> r_ϕ of ϕ by

$$(3.3)\quad 0 \leqq r_\phi = \sum_{x:\lambda(x)=0} \operatorname{ord}_x(\lambda) = \langle c_1(K_M \otimes G), [M]\rangle = 2g - 2 + \langle c_1(G), [M]\rangle .$$

Now let $L \cong \mathbb{C}$ denote the basic representation of the subgroup $U(1)$ of $Sp(n+1)$ in (3.1), and let us retain L to also denote the associated vector bundle on $Z(= \mathbb{C}P^{2n+1})$, which also happens to be identifiable with the hyperplane section bundle and $(T'Z)^\vee \cong L^2$. Our intention is to generalise the notion of the twistor degree in in [8] from $n = 1$, to the conformal harmonic maps arising from the composition

(3.4) $$M_g \xrightarrow{\psi} (Z, J_2) \xrightarrow{q} \mathbb{H}P^n$$

where ψ is the J_2-holomorphic curve of Theorem 3.1 ; we define <u>the twistor degree</u> d of ϕ by setting $d = \langle\psi^*\ell, [M]\rangle$ where $\ell = c_1(L)$.

Recalling the holomorphic line bundle G, we have

(3.5) $$c_1(G) = c_1(\phi^{-1}T'N) - c_1(Q).$$

where $Q = \phi^{-1}T'N/G$ and thus we define <u>the normal characteristic</u> e of ϕ by $e = \langle c_1(Q), [M]\rangle$. In the appendix we shall show that $c_1(Z, J_2) = 2(n-1)\ell$, and on pulling back by ψ, we see that $\psi^*(2(n-1)\ell) = -\psi^*(2\ell) + c_1(\psi^{-1}(T'Z)^h)$. Furthermore, $\psi^{-1}((T'Z)^h) \cong \phi^{-1}T'N$. whence we deduce that

(3.6) $$c_1(\phi^{-1}T'N) \cong \psi^*(2n\ell).$$

Now from (3.3), we have $r_\phi = 2g - 2 + \langle c_1(\phi^{-1}T'N) - c_1(Q), [M]\rangle$ and thus on summing up we have proved:

<u>PROPOSITION 3.2</u> (cf. [8] for the case $n = 1$) <u>The twistor degree</u> d <u>of a non-constant, inclusive conformal harmonic map</u> $\phi : M_g \to \mathbb{H}P^n$, <u>satisfies the relationship</u>

(3.7) $$2nd = r_\phi + 2 - 2g + e.$$

4. EXAMPLES OF HARMONIC MAPS $S^2 \to G_k(\mathbb{H}^{n+k})$.

In this section, we consider the twistor fibration

(4.1) $$\pi: S_{k,n} \cong Sp(n+k)/Sp(n) \times U(k) \to G_k(\mathbb{H}^{n+k}) \cong Sp(n+k)/Sp(n) \times Sp(k)$$

with fibre $Sp(k)/U(k)$, obtained from (2.5) by interchanging k

and n. Firstly, some notation: let us set

$$(4.2)\qquad m^{1,0} = (T'S_{k,n})^h \qquad\qquad n^{1,0} = (T'S_{k,n})^v$$

with $m^{0,1} = \overline{m^{1,0}}$ and $n^{0,1} = \overline{n^{1,0}}$. To exemplify matters, note that

$$(4.3)\qquad \begin{cases} (T'S_{k,n}, J_1) \cong n^{1,0} \oplus m^{1,0} & \text{and} \\ (T'S_{k,n}, J_2) \cong n^{0,1} \oplus m^{1,0} \end{cases}$$

Let $E \cong \mathbb{C}^{2n}$, $G = \mathbb{C}^{2k}$ and $K \cong \mathbb{C}^k$ denote the respective basic representations of the subgroups $Sp(n)$, $Sp(k)$ and $U(k)$ of $Sp(n+k)$ in (4.1), for which we have $G \cong K \oplus \overline{K}$. On taking $S^r(\;)$ to denote the rth symmetric tensor product, we have the Lie algebra identities

$$(4.4)\qquad \mathfrak{sp}(n+k) \cong S^2(E \oplus G),\ \mathfrak{sp}(n) \cong S^2(E),\ \mathfrak{u}(k) \cong K \otimes \overline{K}$$

(we assume the representations complexified), and hence deduce that (pointwise):

$$(4.5)\qquad \begin{cases} m^{1,0} \cong E \otimes K \\ n^{1,0} \cong S^2(K)\,. \end{cases}$$

Recalling the identification $\mathbb{H} \cong \mathbb{C}^2$, we see that the group $Sp(1)$ of unit quaternions acts naturally on $\mathbb{C}^2$ and on setting $p = 2n + 2k - 1$, it acts likewise on $S^p(\mathbb{C}^2) \cong$ {homogeneous polynomials of degree p in two complex variables} $\cong \mathbb{C}^{2n+2k} \cong \mathbb{H}^{n+k}$. This latter representation is quaternionic in so far that $Sp(1)$ commutes with an anti-linear endomorphism $\otimes^p j$ (see e.g. [1]). We thus obtain an inclusion

$$(4.6)\qquad Sp(1) \hookrightarrow Sp(n+k)\,.$$

Observing the identifications $\mathbb{C}P^1 \cong S^2 \cong Sp(1)/U(1)$, we take $\eta \cong \mathbb{C}$ to denote the basic representation of the subgroup $U(1)$ in $Sp(1)$. Furthermore, from the (complexified) Lie algebra decomposition $\mathfrak{sp}(1) = \mathfrak{u}(1) \oplus (\eta^2 \oplus \bar{\eta}^2)$ we have $T'\mathbb{C}P^1 \cong \eta^2$ at the identity coset. Taking the pth symmetric power, we also have

$$(4.7) \qquad S^p(\eta \oplus \bar{\eta}) = \eta^p \oplus \eta^{p-2} \oplus \dots \oplus \eta \oplus \bar{\eta} \oplus \dots \oplus \bar{\eta}^{p-2} \oplus \bar{\eta}^p$$
$$\cong E \oplus G \ (\cong E \oplus K \oplus \bar{K}) .$$

The following assignments (listed A to D below) may be seen to define inclusions

$$(4.8) \qquad \Psi : U(1) \hookrightarrow Sp(n+k)$$

and relative to the natural inclusion $i: U(1) \hookrightarrow Sp(1)$, the basic representation of $Sp(1)$ on $\mathbb{C}^2$ decomposes as $\eta \oplus \bar{\eta}$, and the inclusion (4.6) relative to the inclusions (4.8) induce homogeneous polynomial maps $\psi_R^{(k)}: S^2 \to \mathcal{S}_{k,n}$, where R denotes some indexing set indicating the powers of η arising in the various assignments for K, and hence E (by virtue of (4.7)).

It is therefore convenient for us to denote each such assignment by K_R, from which the corresponding E_R is deduced from (4.7). The resulting properties of the maps $\psi_R^{(k)}$ are determined by applying the relationship in (4.5) and then checking which of $m^{1,0}$, $n^{1,0}$ etc., or any combination of these, contains η^2 (special cases of this construction appear in [8] and [14] to which we refer for further details). One word about terminology: we shall simply say 'horizontal' when we mean that a map is horizontal with respect to π in (4.1).

<u>Assignment A.</u>

Let $A = \{1, 3, \ldots, 2k-1\}$. Define $\psi_A^{(k)} : S^2 \to \mathcal{S}_{k,n}$ by the assignment $K_A = \bigoplus_{i=0}^{k-1} \eta^{a_i}$ where $a_i = 2i + 1$. Then $\psi_A^{(k)}$ is non-horizontal and J_2-antiholomorphic.

<u>Assignment B</u> $(k \geq 2)$

Let $B = \{(b_1, \ldots, b_{k-1}) : b_j$ is an odd integer, $b_1 < b_2 < \ldots < b_{k-1}$ and $a_j \leq b_j \leq 2n + a_j$ for $1 \leq j \leq k-1\} - A$.
Define $\psi_B^{(k)} : S^2 \to \mathcal{S}_{k,n}$ by the assignment $K_B = \eta \oplus \left(\bigoplus_{j=1}^{k-1} \eta^{b_j}\right)$. Then $\psi_B^{(k)}$ is non-horizontal and neither holomorphic nor antiholomorphic.

<u>Assignment C</u>

Let $C = \{ 2n+1, \ldots, p \}$. Define $\psi_C^{(k)} : S^2 \to \mathcal{S}_{k,n}$ by the assignment $K_C = \bigoplus_{i=2n+1}^{p} \eta^i$. Then $\psi_C^{(k)}$ is horizontal and holomorphic. The map $\psi_C^{(1)}$ is, in fact, the Veronese curve $\mathbb{C}P^1 \to \mathbb{C}P^{2n+1}$.

<u>Assignment D</u>

Let $D = \{(d_1, \ldots, d_k) : d_j$ is an odd integer, $d_1 < \ldots < d_k$ and $a_j \leq d_j \leq 2n + a_j - 2$ for $1 \leq j \leq k\} - C$. Define $\psi_D^{(k)} : S^2 \to \mathcal{S}_{k,n}$ by the assignment $K_D = \bigoplus_{i=1}^{k} \eta^{d_i}$. Then $\psi_D^{(k)}$ is horizontal but neither holomorphic nor antiholomorphic.

Other assignments are obtained by replacing η and η^{-1} throughout; we shall omit the details. In all cases the assignments give rise to minimal S^2's in $\mathcal{S}_{k,n}$. In particular, the maps defined by assignment A project under (4.1) to define harmonic maps $S^2 \to G_k(\mathbb{H}^{n+k})$, and in the case of C, we find examples of the isotropic harmonic maps discussed in section 2 (as projections of holomorphic curves from the twistor space $\mathcal{S}_{k,n}$, these are known as 'superminimal immersions' in [4]).

APPENDIX. SOME TOPOLOGY OF THE SPACE $(\mathbb{C}P^{2n+1}, J_2)$.

Recall that $R_{n,1} \cong \mathbb{C}P^{2n+1}$ and from (4.5) we deduce that

$$(T'\mathbb{C}P^{2n+1}, J_2) \cong \bar{L}^2 \oplus (E \otimes L) \tag{A.1}$$

where L is defined as in section 3. Now applying the Whitney product formula, the total Chern class of (A.1) is

$$c(E\otimes L)c(\bar{L}^2) = \left[1+c_1(E\otimes L)+\ldots+c_{2n}(E\otimes L)\right](1+c_1(\bar{L}^2)) \tag{A.2}$$

Thus we deduce that $c_1(\mathbb{C}P^{2n+1}, J_2) = c_1(\bar{L}^2) + c_1(E \otimes L)$.
Now letting $\ell = c_1(L)$, we have $c_1(E \otimes L) = (\text{rank } E)\ell = 2n\ell$ (see [15]), and thus

$$c_1(\mathbb{C}P^{2n+1}, J_2) = 2(n-1)\ell . \tag{A.3}$$

<u>The Todd genus of $(\mathbb{C}P^{2n+1}, J_2)$ is zero.</u>

The result we are about to discuss can in fact be deduced from a result in [3, p. 327]; however, we believe that the proof offered here is a more instructive one. We refer to [12] for an account of the theory of multiplicative sequences and Todd polynomials. We consider then, the multiplicative sequence $\{T_j\}$ of Todd polynomials, and for any almost complex manifold Z, $\dim_{\mathbb{C}} Z = 2n+1$, we define the <u>Todd genus</u> $Td(Z)$ by

$$Td(Z) = \langle T_{2n+1}(c_1, \ldots, c_{2n+1}), [Z]\rangle \tag{A.4}$$

where $T_{2n+1}(\ )$ is the $(2n+1)^{th}$-Todd polynomial (homogeneous of degree 2n+1) in the Chern classes $c_1, \ldots, c_{2n+1}$ of Z.
Recall now the twistor fibration (3.1) $q : \mathbb{C}P^{2n+1} \to \mathbb{H}P^n$ along with the (globally defined) bundle $L \to \mathbb{C}P^{2n+1}$ (the hyperplane section bundle) and $E \to \mathbb{H}P^n$, the bundle associated to the basic

representation of $Sp(n)$ on $\mathbb{C}^{2n}$. Let $Z_1 = (\mathbb{C}P^{2n+1}, J_1)$ and $Z_2 = (\mathbb{C}P^{2n+1}, J_2)$; it is well known that $Td(Z_1) = 1$ (see [12]). The total Chern class of Z_1 is given by

$$\begin{aligned} c(Z_1) &= c\left(L^2 \oplus (L \otimes E)\right) \\ &= c(L^2)c(L\otimes E) = (1+2\ell)c(L\otimes E) \\ &= (1+\ell)^{2n+2} \bmod \ell^{2n+2} \end{aligned}$$

where we recall that $\ell = c_1(L)$. On the other hand,

$$\text{(A.5)} \qquad \begin{aligned} c(Z_2) &= c\left(\bar{L}^2 \oplus (L\otimes E)\right) \\ &= (1-2\ell)c(L\otimes E) \\ &= \left(\frac{1-2\ell}{1+2\ell}\right)(1+\ell)^{2n+2} \bmod \ell^{2n+2} \end{aligned}$$

Now let $T(1+z)$ denote the characteristic power series [11] of the $\{T_j\}$; then we have $T(1+z) = Q(z)$ where $Q(z) = z/1-e^{-z}$. On applying the basic multiplicative properties of the $\{T_j\}$, we see that

$$\begin{aligned} T(1+c_1+\dots+c_{2n+1}) &= T\left(\frac{1-2\ell}{1+2\ell}\cdot(1+\ell)^{2n+2}\right) \\ = \quad & T(1-2\ell)T\left((1+2\ell)^{-1}\right)T\left((1+\ell)^{2n+2}\right) \\ = \quad & T(1-2\ell)\left(T(1+2\ell)\right)^{-1}\left(T(1+\ell)\right)^{2n+2} \end{aligned}$$

which equals

$$\text{(A.6)} \qquad \frac{-2\ell}{1-e^{2\ell}} \quad \frac{1-e^{-2\ell}}{2\ell} \left(\frac{\ell}{1-e^{-\ell}}\right)^{2n+2}$$

Replacing ℓ by the complex variable z, the general theory of [11] tells us that (A.6) is equal to

$$\sum_{j=0}^{m} T_j(c_1, \ldots, c_j)z^j + \sum_{j=m+1}^{\infty} T_j(c_1, \ldots, c_j, 0, \ldots, 0)z^j .$$

Thus the coefficient F_{2n+1} say, of z^{2n+1} in

$$\text{(A.7)} \qquad \frac{-2z}{1 - e^{2z}} \quad \frac{1 - e^{-2z}}{2z} \left(\frac{z}{1 - e^{-z}}\right)^{2n+2}$$

will determine $T_{2n+1}(c_1, \ldots, c_{2n+1})$.

We now apply the Cauchy integral formula and integrate about a small disc centred at the origin:

$$F_{2n+1} = \frac{1}{2\pi i} \oint \frac{1}{z^{2n+2}} \text{(A.7) } dz$$

$$= \frac{1}{2\pi i} \oint - \frac{(1 - e^{-2z})}{(1 - e^{2z})} \cdot \frac{1}{(1 - e^{-z})^{2n+2}} \, dz$$

Using the substitution $u = 1 - e^{-z}$, the above integral reduces to $\frac{1}{2\pi i} \oint \frac{1 - u}{u^{2n+2}} \, du = 0$. Hence we deduce that

$$Td(Z_2) = \langle T_{2n+1}(c_1, \ldots, c_{2n+1}), [Z_2] \rangle = 0 .$$

ACKNOWLEDGEMENT. As the Solomon Lefschetz visiting professor for 1984/85, the author enjoyed the hospitality of the Departamento de Matemáticas, CINVESTAV-IPN, during which time this article was prepared.

BIBLIOGRAPHY

1. J.F. Adams, Lectures on Lie Groups, W.A. Benjamin, New York (1969).

2. A.R. Aithal, 'Harmonic maps from S^2 to $\mathbb{H}P^2$', pre-print Tata Institute (1984).

3. A. Borel and F. Hirzebruch, 'Characteristic classes and homogeneous spaces II', Am. J. Math. 81 (1959), 315-382.

4. R.L. Bryant, 'Lie Groups and twistor spaces', Duke Math, J. Vol. 52, No. 1 (1985) 223-261.

5. J. Eells and L. Lemaire, 'A report on harmonic maps', Bull. London Math. Soc. 10 (1978), 1-68.

6. J. Eells and J.C. Wood, 'Harmonic maps from Riemann surfaces to complex projective spaces', Adv. in Math. Vol. 49 No. 3 (1983), 217-263.

7. J. Eells and S. Salamon, 'Constructions twistorielles des applications harmoniques', C.R. Acad. Sci. Paris 296 (1983), 685-687.

8. J. Eells and S. Salamon, 'Twistorial construction of harmonic maps of surfaces into four-manifolds', Ann. Scuola Norm. Pisa, to appear.

9. S. Erdem and J.C. Wood, 'On the construction of harmonic maps into Grassmannians', J. London Math. Soc. (2), 28 (1983), 161-174.

10. J.F. Glazebrook, 'The construction of a class of harmonic maps to quaternionic projective space', J. London Math. Soc. (2), 30(1984), 151-159.

11. J.F. Glazebrook, 'Isotropic harmonic maps to Kähler manifolds and related properties', Ph.D. Thesis, University of Warwick (1984).

12. F. Hirzebruch, Topological Methods in Algebraic Geometry, Grundlehren 131, 3rd Ed., Springer-Verlag, New York 1966.

13. J.H. Rawnsley, 'f-structures, f-twistor spaces and harmonic maps', Seminar Luigi Bianchi, Springer Lecture Notes, to appear.

14. S. Salamon, 'Harmonic and holomorphic maps', Seminar Luigi Bianchi, Springer Lecture Notes, to appear.

15. S. Salamon, 'Quaternionic Kähler manifolds', Inv. Math. 67 (1982), 143-171.

16. A. Zandi, 'Quaternionic Kähler manifolds and their twistor spaces', University of California-Berkeley, Thesis (1984).

DEPARTAMENTO DE MATEMÁTICAS
CINVESTAV-IPN
APARTADO POSTAL 14-740
MEXICO CP07000, 14, D.F.
MEXICO.

Current Address:
DEPARTMENT OF MATHEMATICS
HILL CENTER
RUTGERS UNIVERSITY
NEW BRUNSWICK, NJ 08903

Contemporary Mathematics
Volume 58, Part III, 1987

FOLIATIONS BY CURVES OF COMPLEX ANALYTIC SPACES*

XAVIER GÓMEZ-MONT

ABSTRACT. Starting from a geometric definition of a foliation by curves on a complex analytic space we derive an equivalent analytic definition: there exists a meromorphic line bundle and a meromorphic map on the smooth points of the space into its tangent bundle whose image are the tangent vectors to the foliation. This map is unique up to multiplication by a meromorphic function. In case the variety is factorial, the line bundle, the map and the uniqueness are in the holomorphic category. In case the variety is normal, there is a local obstruction for the objects to be holomorphic. For the case of normal surfaces we describe a finite group that carries the topological obstructions and a connected analytic group variety that carries the analytic obstructions. Both groups may be read off from a desingularization of the surface.

0. INTRODUCTION.

In the present paper we analyse foliations with singularities by one dimensional leaves on complex analytic spaces with singularities. We define the foliations as non-singular foliations outside of a codimension two subvariety. This restriction seems natural since if we do not specify the foliation in a codimension one subvariety, it could have an essential singularity along it. On the other hand, we have extension theorems for analytic objects to subvarieties of codimension bigger than one, and one of the purposes of the present paper is to persue this extension procedure and see what kind of objects we obtain.

For non-singular spaces in codimension one we observe that the tangent vectors to the foliation form in a natural way a meromorphic line bundle. This comes from the fact that the foliation may always be defined locally by a meromorphic vector field that is unique up to multiplication by a non-identically zero meromorphic function . It is from the non-uniqueness that we obtain the meromorphic line bundle.

If the variety is factorial (i.e. all the local rings are unique factorization domains) then we may reduce the structure group of the bundle to obtain a holomorphic bundle. This is a caused by the fact that the ambiguity of specifying a

*Supported by CONACYT and CNPq at IMPA.

© 1987 American Mathematical Society
0271-4132/87 $1.00 + $.25 per page

local holomorphic vector field tangent to the foliation that vanishes on a set of codimension bigger than one is by multiplying by a never zero holomorphic function.

In case the variety is just normal (i.e. all the local rings are integrally closed) there is a local obstruction to reducing the structure group of the line bundle. This is caused by the fact that we may not always specify locally a foliation by a holomorphic vector field vanishing outside of a codimension one subvariety, but in case that we can then it is unique up to multiplication by a never vanishing holomorphic function.

We exhibit explicit examples of normal surfaces where this obstruction appears. Working on normal surfaces we are able to detect a finite group, whose rank is the determinant of the intersection pairing of a desingularization of the normal surface, that measures the topological obstructions, and a connected analytic group variety, whose dimension is the irregularity of a desingularization of the surface, that measures the analytic obstructions. We rely heavily on Mumford's description of curves on a singular surface [8] to derive these last results.

Foliations in singular surfaces are an intermediate step between foliations by curves in two and three dimensions, and seem to be a natural place where one can begin to extend the known classical theorems on differential equations in two variables (see Camacho [1]). Laying the proper definitions for this project was the original motivation of this work. Preliminary versions of this paper have appeared in [3] and [4].

The paper has three sections. In the first we sketch the results we will need later on, centering on analytic spaces, line bundles and vector fields. In the second section we prove the local theorem 5, from which we obtain the extensions and the non-uniqueness. In this section we also show how to glue the local data and give a series of global applications. In the last section we concentrate on foliations in surfaces.

1. TWISTED MEROMORPHIC VECTOR FIELDS.

In this section we will briefly recall the basic notions of complex analytic geometry that we will use later on, centering on analytic spaces, line bundles and vector fields. More details may be found in Fischer [2], Griffiths-Harris [5], Gunning-Rossi [6], Narasimhan [9] and Rossi [10].

A complex analytic space is a topological space where we have specified a ring of 'holomorphic' functions and that we construct by giving some natural local models. The local models are subsets of open balls in complex number spaces $B = \{z \in \mathbb{C}^N \mid \|z\| < r\}$ defined by a finite number of complex analytic equations:

$$W = \{z \in B \subset \mathbb{C}^N \mid f_k(z) = 0,\ k = 1,\dots,n\} \qquad (1)$$

and the holomorphic functions on W are the holomorphic functions of B restricted to W.

Formally, an *analytic space* is a Hausdorff topological space V provided with a countable number of open subsets V_i and bicontinuous maps $\varphi_i: V_i \to W_i \subset B \subset \mathbb{C}^N$ as in (1) such that the transitions coordinates $\varphi_j \circ \varphi_i^{-1}: \varphi_i(V_i \cap V_j) \to \varphi_j(V_i \cap V_j)$ are biholomorphic maps between open pieces of the local models. A complex valued function $f: V \to \mathbb{C}$ is said to be *holomorphic* if for every $i, f \circ \varphi_i: W \to \mathbb{C}$ is a holomorphic function on W_i. An *analytic subvariety* A of V is a subset of V such that $\varphi_i(A \cap V_i) \subset W_i \subset B_i$ is defined as the zero set of some holomorphic functions of B_i.

A point p in V is *smooth* if we may find n holomorphic functions $f_1, \ldots, f_n$ defined in a neighbourhood V' of p in V that define a homeomorphism of V' onto an open subset of C^n. If a point is not smooth then it is *singular*. The set of singular points of V is denoted by *Sing* V and it is a complex analytic subvariety of V. V-sing V is a complex manifold and its (complex) dimension is the biggest n of the range $\mathbb{C}^n$ of its smooth coordinate charts. The *dimension of* V *at* p is

$$\dim_p V = \max\{\dim(U\text{-Sing}\, U) \,/\, U \text{ neighbourhood of } p\}$$

the *dimension of* V is the dimension of V-Sing V.

Analytic space V is *irreducible* if it cannot be written as the union of two analytic subvarieties V_1, V_2 with $V_1, V_2 \neq V$. V is *irreducible at* p if we may find a basis of neighbourhood U_i of p where $U_i \cap V$ is irreducible. Note that by definition all our analytic spaces are reduced.

An analytic space is of *pure dimension* n if all the connected components of V-Sing V have dimension n. An analytic space of pure dimension n is said to be *non-singular in codimension* 1 if $\dim \text{Sing}\, V < n-1$.

The *local ring at* p *in* V, $\mathcal{O}_{V,p}$, is formed by the holomorphic functions defined in a neighbourhood of p, where we identify two if they coincide in a neighbourhood of p, i.e. germs of holomorphic functions at p.

V is *factorial at* p if the local ring $\mathcal{O}_{V,p}$ is a unique factorization domain. V is *factorial* if it is factorial at every point of V.

V is *normal at* p if the local ring $\mathcal{O}_{V,p}$ is an integrally closed ring. V is *normal* if it is normal at every point of V.

In the following proposition we collect the main facts that we will use from the above definitions:

PROPOSITION 1: 1) *If* V *is smooth at* p *then it is also factorial at* p.

2) *If* V *is factorial at* p, *then it is normal at* p.

3) *If* V *is normal at* p, *then it is irreducible at* p *and in a neighbourhood of* p *it is non-singular in codimension* 1.

4) *A normal variety* V *is factorial at* p *if and only if every subvariety through* p *of pure dimension* $\dim_p V-1$ *(with multiplicities) may be defined in a neighbourhood of* p *as the zero set of one holomorphic function.*

A proof of this proposition may be found in the references, and part 4 in section II.6 of Hartshorne [7].

The condition that V is locally irreducible at p may be expressed also by the fact that the local ring $\mathcal{O}_{V,p}$ has no divisors of zero, i.e. it is a domain. But then we may form the algebraic field associated with the domain to obtain the *germs at* p *of meromorphic functions on* V.

$$M_{V,p} = \{f/g \mid f,g \in \mathcal{O}_{V,p}\ ,\ g \neq 0\}$$

Of course these are not functions in the usual sense, since it has poles where $g = 0$ and is undetermined where $f = g = 0$. A *meromorphic function* on an open set U of V is a "function" that belongs to the fields $M_{V,p}$ for every p in U, or more formally, it is a section of the sheaf obtained from the stalks $M_{V,p}$. To avoid too much technicalities, a meromorphic function is a function that locally is expressed as the quotient of two holomorphic functions and we will only consider them when the space is locally irreducible at each point.

We will need to use extension theorem of Hartog's Type:

<u>Riemann's Removable Singularity Theorem</u>: *Let* V *be a normal complex space and* $A \subset V$ *an analytic subset of codimension at least 2 in every point. Then every holomorphic function on* V-A *has a unique holomorphic extension to* V.

<u>Levi's Removable Singularity Theorem</u>: *Let* V *be a locally irreducible complex space with an analytic subset* A *of codimension at least 2 in every point. Then every meromorphic function on* V-A *has a unique extension to a meromorphic function on* V.

Proofs of these theorems may be found in Fischer [2], pp. 120-185.

The main tool we will need for what follows is the notion of line bundles. Intuitively a line bundle is a holomorphic family of complex vector spaces of dimension 1 parametrized by some variety V. Formally, a *holomorphic line bundle* over the analytic space V may be described by an open cover $\{V_i\}$ of V and non-vanishing holomorphic functions $\xi_{ij}: V_i \cap V_j \to \mathbb{C}-\{0\} = \mathbb{C}^*$ defined on $V_i \cap V_j$ and satisfying the cocycle condition $\xi_{ij} \cdot \xi_{jk} = \xi_{ik}$ in $V_i \cap V_j \cap V_k$. To construct the line bundle from the data (ξ_{ij}) we consider $V_i \times \mathbb{C}$ and glue with $V_j \times \mathbb{C}$ along $(V_i \cap V_j) \times \mathbb{C}$ through the map $(V_i \cap V_j) \times \mathbb{C} \to (V_i \cap V_j) \times \mathbb{C}$,

$(z,t) \to (z,\xi_{ij}(z)\cdot t)$.

The compatibility of the gluing follows from the cochain condition and since we are giving each fiber by means of a $\mathbb{C}$-linear transformation, each fiber inherits a natural structure of a 1-dimensional vector space, $\pi: L \to V$. Two line bundles L and L' over V are biholomorphically equivalent if we may find a biholomorphism $\varphi: L \to L'$ over V that is $\mathbb{C}$-linear over each fiber. In terms of the cocycle (ξ_{ij}) and (ξ'_{ij}) it means that after refining the covering if necessary $\xi'_{ij}\cdot\xi_{ij}^{-1} = \eta_i\cdot\eta_j^{-1}$ for some η_i: $U_i \to \mathbb{C}^*$. So it is completely classified by $H^1(V,\mathcal{O}_V^*)$ where we are using Cech-cohomology with coefficients in the multiplicative sheaf of never vanishing holomorphic functions.

In analogy, a *meromorphic line bundle* on the locally irreducible space V may be described by an open covering $\{V_i\}$ of V and non-identically equal to zero meromorphic functions ξ_{ij} defined on $V_i \cap V_j$ and satisfying the cocycle condition $\xi_{ij}\cdot\xi_{jk} = \xi_{ik}$ on $V_i \cap V_j \cap V_k$. There is also a notion of equivalence of meromorphic line bundles (bimeromorphic equivalence) which may be expressed by saying that (ξ_{ij}) is an element of $H^1(V,M_V^*)$ where we are using Cech-cohomology with coefficients in the multiplicative sheaf of non-identically equal to zero meromorphic functions M_V^*.

A *holomorphic (or meromorphic) map between* two holomorphic (or meromorphic) *line bundles* $\xi = (\xi_{ij})$ and $\eta = (\eta_{ij})$ may be described by a set of holomorphic (or meromorphic) functions (φ_i) defined on the open sets V_i and satisfying the compatibility conditions $\eta_{ij}\circ\varphi_i = \varphi_j\circ\xi_{ij}$ or what is the same $\varphi_i = (\xi_{ij}\,\eta_{ij}^{-1})\cdot\varphi_j$ that is expressed by saying that φ_i is a global section of the line bundle that has transition functions $(\xi_{ij}\,\eta_{ij}^{-1}) = \xi \otimes \eta^{-1}$.

Every holomorphic line bundle gives rise to an associated meromorphic line bundle, since $\mathcal{O}_V^* \subset M_V^*$. The line bundle whose transition data $\xi_{ij} = 1$ will be called the *trivial line bundle*, 1_V.

Any meromorphic map of the trivial line bundle onto itself is then described by a family of meromorphic functions φ_i defined on V_i and satisfying $1\cdot\varphi_i = \varphi_j\cdot 1$. So they coincide on the intersection and give rise to a meromorphic function on V. More generally

LEMMA 2: *Holomorphic (or meromorphic) maps from any holomorphic (or meromorphic) line bundle into itself correspond to holomorphic (or meromorphic) functions.*

Proof: These maps then are given by a family of functions (φ_i) satisfying $\xi_{ij}\varphi_i = \varphi_j\xi_{ij}$ on $U_i \cap U_j$. where (ξ_{ij}) are the coordinate transition functions of the bundle. So cancelling ξ_{ij}, they give rise to a globally defined function $\varphi_i = \varphi_j$ on all or V.

Let V be a complex analytic space. To define a germ of a holomorphic vector field at $p \in V$, let $i: V \to C^N$ be a germ of an embedding of V at p into C^N, and assume that the ideal of holomorphic functions vanishing on $i(V) = V$ at $0 = i(p)$ is generated by $f_1,\ldots,f_r$. We say that a germ of a holomorphic vector field $(X_1,\ldots,X_N)$ of C^N is tangent to V if for $i = 1,\ldots,r$

$$\sum_{j=1}^{N} \frac{\partial f_i}{\partial z_j} X_j \in (f_1,\ldots,f_r) \subset \mathcal{O}_{\mathbb{C}^N,0}$$

where $(f_1,\ldots,f_r)$ denotes the ideal generated by $f_1,\ldots,f_r$ in the local ring of holomorphic functions at 0 in C^N. $(X_1,\ldots,X_N)$ and $(Y_1,\ldots,Y_N)$ tangent to V define the same vector field on V (or are equivalent) if and only if for $j = 1,\ldots,N$ we have that $X_j - Y_j \in (f_1,\ldots,f_r)$. We then define a *germ of a holomorphic vector field on* V *at* p by an equivalence class of holomorphic vector fields tangent to V.

We shall denote by $\Theta_{V,p}$ the germs of holomorphic vector fields on V at p. If V is an analytic space, a holomorphic vector field X will be a section of the sheaf $\Theta_{V,p}$, i.e. the germ of X at p will be in $\Theta_{V,p}$ for all p in V. For such a vector field X, we will denote by $(X)_0$ *the set of points where* X *vanishes*. It is clear that $(X)_0$ is an analytic subvariety of V at p. The proof of the following theorem may be found in Rossi [10]:

THEOREM 3: *Let* V *be a complex analytic space and* p *a point of* V, *then:*

1) *The definition of a germ of a holomorphic vector field is independent of the embedding of* V *into* C^N.

2) *The sheaf of holomorphic vector fields* Θ_V *is a coherent analytic sheaf over* $\mathcal{O}_V$.

In case V is a complex manifold of dimension n, we have associated to it the *tangent bundle of* V T_V. It is a holomorphic n-dimensional vector bundle that may be given by transition coordinates the Jacobians $(D\varphi_{ij})$ of a covering by coordinate charts $\{(V_i,\varphi_i)\}$. Let ξ be a holomorphic line bundle over V, given by the transition coordinates (ξ_{ij}) over the same covering $\{V_i\}$. A *holomorphic map of the line bundle* ξ *to the tangent bundle* may be given by a family of holomorphic maps

$$V_i \times \mathbb{C} \to V_i \times \mathbb{C}^n,\ (z,t) \to (z,\alpha_i(z)\cdot t)$$

over V_i linear over each fiber such that on $V_i \cap V_j$

$$\alpha_i \cdot \xi_{ij} = D\varphi_{ij} \cdot \alpha_i \tag{2}$$

that is, the following diagram commutes

$$\begin{array}{ccc} V_i \times \mathbb{C} & \xrightarrow{\alpha_i} & V_i \times \mathbb{C}^n \\ \xi_{ij} \downarrow & & \downarrow D\varphi_{ij} \\ V_j \times \mathbb{C} & \xrightarrow[\alpha_j]{} & V_j \times \mathbb{C}^n \end{array}$$

This is the notion that generalizes a vector field in a complex manifold, since the α_i may be seen as local vector fields, but we are including the fact that globally the α_i may not glue to give a trivial bundle, as is the case for holomorphic vector fields.

If the maps $\alpha_i = (\alpha_i^1,\ldots,\alpha_i^n)$ are now meromorphic but still satisfy (2), we will say that they glue to give a *meromorphic map from the holomorphic or meromorphic line bundle* ξ *to the tangent bundle*. In particular, if ξ is the trivial bundle, $\xi_{ij} = 1$, this gives the definition of a *meromorphic vector field on the smooth manifold* V.

We would like now to define meromorphic vector fields on the locally irreducible analytic space V. If V is locally embedded in C^N, with $f_1,\ldots,f_r$ generating the ideal of holomorphic functions on C^N vanishing on V, a *germ of a meromorphic vector field* will be an N-Tuple $(X_1,\ldots,X_N)$ where the X_i are meromorphic functions on V that they satisfy for $i = 1,\ldots,r$

$$\sum_{j=1}^{N} \frac{\partial f_i}{\partial z_j} X_j = 0 \qquad \text{on } V$$

For later use we record the

LEMMA 4: *If* X *is a meromorphic function or meromorphic vector field that vanishes whenever defined, then it is the zero meromorphic function or vector field.*

2. FOLIATIONS BY CURVES.

In this section we begin by defining the main object under study, foliations by curves of complex analytic spaces. We will concentrate first in the local problem of analyzing when the foliation can be described around an arbitrary point by a vector field that is tangent to the foliation, and how unique can we choose this vector field. From the local non-uniqueness we will associate to a global foliation a meromorphic line bundle, and in case the variety is factorial, a holomorphic line bundle. These bundles are interesting invariants of the foliation.

Definitions: A *non-singular foliation by curves* F of the complex manifold M may be defined by an open covering of coordinate charts $\{(V_i,\varphi_i)\}$ such that the transition functions $\varphi_i \circ \varphi_j^{-1} = \varphi_{ij} = \varphi = (\varphi^1,\ldots,\varphi^n)$ satisfy $\frac{\partial\varphi^k}{\partial z_n} = 0$ for $k = 1,\ldots,n-1$.

Let V be a complex analytic space of pure dimension n, non-singular in codimension one. A *foliation by curves* F of V will be a non-singular foliation of the complex manifold $V - W$, where W is a complex subvariety of V of codimension bigger than one that contains the singular set of V, $W \supset \text{Sing } V$.

Let F be a foliation by curves of V, non-singular on $V - W$ Our technique for finding local holomorphic vector fields tangent to F at the points of W will be to use the sub-line bundle of the tangent bundle of the smooth manifold $\tau \to T_{V-W}$ consisting of vectors tangent to the foliation F. A close analisis of this line bundle and the obstructions to extending it as an abstract line bundle to the points of W will give some insight to the relationship between foliations by curves and meromorphic vector fields.

Assume the foliation on $V - W$ is defined by an open covering $\{(V_i,\varphi_i)\}$ such that the transition functions $\varphi = \varphi_{ij} = (\varphi^1,\ldots,\varphi^n)$ satisfy $\frac{\partial\varphi^k}{\partial z_n} = 0$ for $k = 1,\ldots,n-1$. This differential condition has the geometric implication of sending the set of vertical lines $\{(a_1,\ldots,a_{n-1})\times \mathbb{C}\}$ into itself by means of φ. Globally these vertical lines form the leaves of the foliation of V-W. The set of tangent vectors to the leaves will form a subset $\tau \hookrightarrow T_{V-W}$ of the tangent bundle of V-W and by the local form of the foliation it is actually a 1-dimensional line bundle embedded in T_{V-W}. It is trivial $V_i \times C$ over each coordinate chart V_i and it has transition functions $\left(\frac{\partial\varphi^n_{ij}}{\partial z_n}\right)$ on $V_i \cap V_j$. The chain rule and the differential conditions of the definition guarantee the cocycle condition, and hence it defines an abstract holomorphic line bundle on V-W.

The main result of this section is:

THEOREM 5: *Let* V *be a germ of an irreducible complex analytic space at* p, *non-singular in codimension 1 and* F *a foliation by curves of* V, *non-singular on* V-W. *Then:*

1) *We may find a germ of a holomorphic vector field of* V *at* p, $X \not\equiv 0$, *that is tangent to* F *whenever it is non-vanishing.* X *is unique up to multiplication by a meromorphic function at* p.

2) *If* V *is a normal variety and if we may find an* X *with the zero set of* X, $(X)_0$, *of codimension bigger than one, then any other such vector field differs from* X *by multiplication by a non-vanishing holomorphic function.*

3) *If* V *is factorial at* p, *then we may always find* X *with codimension of* $(X)_0$ *bigger than one.*

4) *We may find a finite number* $X_1,\ldots,X_s$ *of holomorphic vector fields tangent to* F *such that for any other holomorphic vector field* X *tangent to* F *we have* $(X)_0 \supset \bigcap_{i=1}^{s} (X_i)_0$.

Proof: 1) Choose a germ of an embedding of V into some C^N with p = 0. Denote by τ the line bundle on V-W tangent to the foliation, with its natural embedding $\tau \hookrightarrow T_{V-W}$. Since V is embedded in C^N we also obtain a natural embedding of the tangent bundle of V-W into the tangent bundle of C^N restricted to V-W, $T_{V-W} \hookrightarrow T_{\mathbb{C}^N}|_{V-W}$. Hence by composition, an embedding $\tau \hookrightarrow T_{\mathbb{C}^N}|_{V-W}$. Since the tangent bundle of C^N is a direct sum of N copies of the trivial bundle, the last embedding may be expressed by $(X_1,\ldots,X_N)$: $\tau \to T_{\mathbb{C}^N}|_{V-W} = \oplus^N 1_{V-W}$ where each X_i is a holomorphic line bundle map $X_i\colon \tau \to 1_{V-W}$ from τ to the trivial bundle on V-W. Assume by reordering the coordinates that X_1 is not identically equal to zero, and let D be the set where X_1 vanishes, $D = \{X_1 = 0\}$. Since X is locally expressed as a function, D is empty or of codimension 1, and its inverse $X_1^{-1}\colon 1_{V-W} \to \tau$ is a meromorphic map with poles along D. Composing with X_j we obtain meromorphic maps $\varphi_j\colon X_j \circ X_1^{-1}\colon 1_{V-W} \to \tau \to 1_{V-W}$ which may be seen by Lemma 2 as given by the meromorphic functions φ_j on $V-W$. Applying Levi's removable singularity theorem for meromorphic functions we obtain that φ_j are actually meromorphic functions on V, whose poles are completely contained in $D \cup W$.

We claim now that $Y = (1,\varphi_1,\ldots,\varphi_N)$ is a meromorphic vector field on V tangent to the foliation F outside of $D \cup W$.

Note that $Y \circ X_1 = (X_1, \varphi_2 \circ X_1, \ldots, \varphi_N \circ X_1) = (X_1, X_2, \ldots, X_N)$ viewed as a composition $\tau \to 1$ and $1 \to T_{V-W}$, so that both have the same image outside $D \cup W$. Hence Y is tangent to the foliation outside of $D \cup W$. Now, if $f_1,\ldots,f_r$ are defining equations of V in C^N, $(df_i)Y$ is a meromorphic function on V that vanishes outside of $D \cup W$, so it must vanish identically on V by Lemma 4. This makes Y into a meromorphic vector field on V tangent to F whenever defined.

Locally at 0 we may express the components of Y as the quotient of holomorphic functions, so multiplying Y by a holomorphic function g that is a multiple of all the denominations, we obtain $X = gY$ a holomorphic vector field tangent to F whenever it is not zero.

If $Z = (Z_1,\ldots,Z_N)$ is another such holomorphic vector field, then $Y - \frac{1}{Z_1} Z$ is a meromorphic vector field that vanishes on the complement of an analytic subset, since both Y and $\frac{1}{Z_1} Z$ point tangent to F and their first coordinate is 1. By Lemma 4 Y is equal to $\frac{1}{Z_1} Z$ so that $X = \frac{g}{Z_1} Z$ so that X and Z differ by multiplication by a meromorphic function at 0. This proves 1).

2) Assume now that X is a normal variety and that we may find two holomorphic vector fields near 0, X_1 and X_2, tangent to F and vanishing on a set of codimension bigger than one, D_1 and D_2. Viewing X_1 and X_2 as maps from the trivial line bundle 1 to the tangent bundle of C^N restricted to V, since they have the same image, we may consider the holomorphic map

$$X_2^{-1} \circ X_1\colon 1_{V-(D_2 \cup D_2)} \to 1_{V-(D_2 \cup D_2)}$$

which we view as a holomorphic function. Now using Levi's extension theorem for holomorphic functions we see that $X_2^{-1} \circ X_1 = h$ where h is a holomorphic function defined in all of V, and hence $X_1 = hX_2$. Since h is a function it vanishes on a set of codimension 1 or it does not vanish at all. But since X_1 vanishes only on a set of codimension bigger than one, then it is actually non-vanishing. This proves 2).

3) If V is factorial at p, consider the meromorphic vector field constructed in 1, and express it locally as the quotient of relatively prime holomorphic functions $(X_1,\ldots,X_N) = (\frac{f_1}{g_1},\ldots,\frac{f_N}{g_N})$.

Clearing out denominators to obtain a holomorphic vector field, and cancelling all common multiples we end up with a vector field that vanishes on a set of codimension bigger than one on V, for otherwise we could still divide by the equation of this subvariety of codimension one, contradicting the construction of cancelling all common multiples.

4) If we consider all local holomorphic vector fields tangent to F at 0, then the germ of the analytic set $\cap\{X=0\}$ is well defined, and by the Noetherian properties of the local ring $\mathcal{O}_{V,0}$, it may be cut out by a finite number $X_1,\ldots,X_s$. □

That the hypothesis of normal and factorial varieties in the statement of the theorem may not be weakened may be seen from the following example:

Example 1: Consider in $\mathbb{C}^2$ the involution $T(Z_1,Z_2) = -(Z_1,Z_2)$, and let $S = \mathbb{C}^2/(T)$ be the quotient space of $\mathbb{C}^2$ by T. S may be embedded in C^3 with equation $Y^2 = XZ$, where it is seen to be a normal non-factorial variety (Hartshorne [6], pp. 133). The involution T also acts on germs of holomorphic functions and vector fields, splitting them into ± 1 eigenspaces

$$\mathcal{O}_{\mathbb{C}^2,0} = \mathcal{O}^+_{\mathbb{C}^2,0} \oplus \mathcal{O}^-_{\mathbb{C}^2,0} \qquad \Theta_{\mathbb{C}^2,0} = \Theta^+_{\mathbb{C}^2,0} \oplus \Theta^-_{\mathbb{C}^2,0}$$

where the invariant functions and vector fields correspond to the function and vector fields on S.

Note that expanding in power series around 0, $\mathcal{O}^+_{\mathbb{C}^2,0}$ and $\Theta^-_{\mathbb{C}^2,0}$ contain only even powers and $\mathcal{O}^-_{\mathbb{C}^2,0}$ and $\Theta^+_{\mathbb{C}^2,0}$ odd powers. In particular, the elements in $\mathcal{O}^-_{\mathbb{C}^2,0}$ and $\Theta^+_{\mathbb{C}^2,0}$ always vanish at 0.

If F is a germ of a foliation by curves on S, and F is the foliation induces by F on $\mathbb{C}$, it is not difficult to prove (see [4]) that F may be defined by a holomorphic vector field X vanishing only at 0, with a dichotomy X in $\Theta^+_{\mathbb{C}^2,0}$ or $\Theta^-_{\mathbb{C}^2,0}$. In the first case $\tilde{X}$ will induce a holomorphic vector on S vanishing only at 0. But in case $\tilde{X}$ is anti-invariant, we need to multiply by an anti-invariant function $g\cdot\tilde{X}$, $g \in \mathcal{O}^-_{\mathbb{C}^2,0}$ to construct an invariant vector field tangent to the foliation. But since g vanishes at 0, it vanishes on a codimension 1 subset. Take for example $\tilde{X} = \frac{\partial}{\partial Z_1}$. We may then see for these foliations that we may not improve statement 3 of the theorem from factorial to normal. We will give a topological justification of this dichotomy in the next section.

Remarks on Theorem 5: 1) On a local level, a foliation by curves on an irreducible, non-singular in codimension 1 space is naturally defined analytically by a meromorphic vector field tangent to the foliation, since the ambiguity of specifying it in two different ways is by multiplication by a meromorphic function. This fact hints that foliations should be considered as meromorphic objects.

2) In case V is factorial at p, then the foliation may be specified naturally by a holomorphic vector field vanishing on a set of codimension bigger than one, since then the ambiguity comes from multiplying by a holomorphic non--vanishing function. These contain the smooth points of V, see [3]. In this situation, foliations can be considered as holomorphic objects.

3) If V is normal at p, then there is a local obstruction to define a foliation as a holomorphic object. We will consider again this point in the next section.

We now show how to use the local Theorem 5 to give analytic expressions for foliations. It will turn out that in non-factorial varieties, foliations are meromorphic objects, while on factorial varieties they may be considered as holomorphic.

THEOREM 6: *Let* V *be a factorial complex analytic variety and* F *a foliation by curves of* V. *There is a uniquely determined holomorphic line bundle* τ *on* V *and a holomorphic map* τ *of* φ *restricted to* $V' = V\text{-Sing}\,V$ *into the tangent bundle of* V', *unique up to multiplication by a never-vanishing holomorphic function on* V, *such that the image of* φ *is the set of tangent vectors to the foliation.*

Proof: By part 3 of Theorem 5 we may find an open cover $\{V_i\}$ of V and holomorphic vector fields X_i on V_i tangent to the foliation F on V_i and vanishing on a set of codimension bigger than one. On $V_i \cap V_j$ we have two such expressions, so by part 2 of the same theorem $X_i = \tau_{ij}X_j$ for some non-vanishing holomorphic functions τ_{ij}. The collection of holomorphic function (τ_{ij}) satisfy the cochain condition of a line bundle by definition. These are the cocycle expressions of the line bundle τ of the theorem. We may view the X_i restricted to $V' \cap V_i = V_i'$ as giving maps of the trivial bundle into the tangent bundle of V_i

$$X_i : 1_{V_i'} \to T_{V_i'}$$

and the defining equation of τ_{ij} guarantee that we may glue them to form a global map $\varphi\colon \tau|_{V'} \to T_{V'}$.

If φ' is another such map, then $\varphi' \circ \varphi^{-1}$ is a map of τ restricted to $V'\text{-Sing}\,F$ onto itself, so by Lemma 2 it corresponds to a holomorphic function on $V\text{-Sing}\,F$, but since by hypothesis $\text{Sing}\,V$ and $\text{Sing}\,F$ have codimension bigger than one, and since factorial varieties are normal, Riemann's removable singularity theorem shows that $\varphi' \circ \varphi^{-1}$ is a holomorphic non-vanishing function on V. □

COROLLARY 7: *If* V *is a normal variety and if locally the foliation on* V *may be prescribed by a holomorphic vector field vanishing on a set of codimension bigger than one the same conclusion of Theorem 6 applies.*

Proof: The same proof works once the local obstruction cancel. □

THEOREM 8: *Let* V *be an analytic space, locally irreducible and non-singular in codimension one, and let* F *be a foliation by curves of* V. *Then there is a uniquely determined meromorphic line bundle* τ *on* V *and a meromorphic map* φ *restricted to* $V' = V\text{-Sing}\,V$ *into the holomorphic tanget bundle* $T_{V'}$, *unique up to multiplication by a non-identically equal to zero meromorphic function on* V, *such that the image of* φ *is the set of tangent vectors to the foliation outside of the zeroes and poles of* φ.

Proof: Mimic the proof of Theorem 6, noting that part 1 of theorem 5 gives the transition coordinates for the meromorphic line bundle. □

If V is a projective variety, i.e. a closed subvariety of some projective space $\mathbb{C}P^N$, then using the relationship between algebraic and analytic geometry, we can say much more:

THEOREM 9: *Let* V *be a projective variety, locally irreducible and non-singular in codimension one:*

1) *There is a one to one correspondance between rational vector fields on* V *(up to multiplication by a rational function) and foliation by curves of* V.

2) *If besides* V *is factorial, then the line bundle associated to the divisor of poles of the rational vector field* X *defining* F *coincides with the holomorphic line bundle that is tangent to the foliation.*

3) *If* V *is factorial, then there is a finite dimensional family of foliations that have the same abstract holomorphic line bundle tangent to the foliation.*

Proof: 1) Follows from Theorem 8 and the observation that on a projective variety all meromorphic line bundles are trivial.

2) Follows by the observation that cancelling the poles of X and the divisor of V where X vanishes, we obtain a holomorphic map $X\colon \tau \to \Theta_V$ and hence τ may be recognized from the analytic class of the poles and zeroes of X.

3) Given τ, the set of all maps of τ into Θ_V is $H^0(V,\mathrm{Hom}(\tau,\Theta_V))$, and by a well known Theorem of Cartan this space is finite dimensional. Since all holomorphic functions on V are constants, projectivize H^0 and a Zarizki open set is the family of all foliations on V with given bundle τ. □

3. FOLIATIONS ON SURFACES.

In this section we analyse the phenomena, brought out from Example 1, of foliations that may not be prescribed by a vector field that vanish on a set of codimension bigger than one. We first formulate this condition in terms of the analytic triviality of certain holomorphic line bundle, problem which then splits first into a topological (is it topologically trivial?) and then an analytic (if it is topologically trivial, is it then analytically trivial?). We pass to the concrete situation of a surface with an isolated singularity and using a result of Mumford we can show that the topological obstructions form a finite group whose rank may be read from the desingularization of the surface, and that then the analytic obstructions form a connected analytic group.

Let V be a germ of a normal analytic space embedded in some $\mathbb{C}^N$ and F a foliation by curves of V. Let $V' = V\text{-}\mathrm{Sing}\,V$ be the complex manifold obtained by deleting the singular points of V. The foliation F restricts to a foliation in V', and since V' is non-singular, there is a holomorphic line bundle τ' defined on V' and a holomorphic bundle map $\varphi\colon \tau' \to T_{\mathbb{C}^N}|_{V'}$ defined uniquely up to multiplication by a non-vanishing holomorphic function on V' whose image are the tangent lines to the foliation, as in Theorem 6. After shrinking perhaps V near 0 we have:

PROPOSITION 10: *There is a germ of a holomorphic vector field on V at 0 tangent to F and vanishing on a set of codimension bigger than one if and only if τ is the holomorphically trivial line bundle on $V' = V\text{-}\mathrm{Sing}\,V$.*

Proof: $\Rightarrow$) Assume there is such a local vector field X. We may view it, after restricting to V' and perhaps shrinking V, as a bundle map

$$X\colon 1_{V'} \to T_{V'}$$

Now τ' comes provided with a map

$$\varphi\colon \tau' \to T_{V'}$$

Both maps have the same image outside of sets of codimension bigger than one, so define

$$\varphi^{-1}\circ X'\colon 1_{V'} \to \tau'$$

It will be an isomorphism outside of a set of codimension bigger than one. By the Riemann removable singularity theorem and the fact that such a map vanishes on a set of codimension one or not at all, we see that it extends as an isomorphism over all of V'. Hence τ' is analytically trivial

$\Leftarrow$) If τ' is analytically trivial, after considering the embedding $T_{V'} \hookrightarrow T_{\mathbb{C}^N}|_{V'}$ gives a map

$$\varphi\colon 1_{V'} = \tau_{V'} \to T_{\mathbb{C}^N}|_{V'}$$

which extends to V as a holomorphic vector field tangent to V by Riemann's removable singularity theorem. □

This proposition points out then that there is a local obstruction on V-Sing V to finding a holomorphic vector field tangent to F and vanishing on a set of codimension bigger than one. We will now restrict to a simple situation, that sheds some light on the general case and that will clarify example 1.

Let S be germ of a normal analytic surface at 0, and F a germ of a foliation by curves. Denote by τ the holomorphic line bundle on $S' = S - \{0\}$ tangent to the leaves of the foliation F restricted to S' (as specified by theorem 6 applied to the smooth surface S'). Let $\tilde{S}$ be a *desingularization of* S at 0, that is $\tilde{S}$ is a smooth complex surface provided with a proper holomorphic map $\pi: \tilde{S} \to S$ such that $\pi^{-1}(0)$ is the union of a finite set of smooth compact Riemann surfaces E_i, if $i \neq j$ and $E_i \cap E_j \neq \phi$ then E_i and E_j intersect transversely in exactly one point, which does not lie on any other E_k and $\pi: \tilde{S} - \cup E_i \to S'$ is a biholomorphism.

LEMMA 11: *The foliation by curves* F *on* S *induces in a natural way a foliation by curves* $\tilde{F}$ *on the desingularization* $\tilde{S}$ *of* S.

Proof: Since $\pi: \tilde{S} - \cup E_i \to S'$ is a biholomorphism, the foliation on S induces a foliation F' on $\tilde{S} - \cup E_i$ by means of π. We have to show that F' extends to all of $\tilde{S}$. To do this, it is enough to show that it extends to the smooth points of $\cup E_i$, since $\bigcup_{i \neq j} (E_i \cap E_j)$ has codimension 2 in S.

Locally on $E_i - \bigcup_{j \neq i} (E_i \cap E_j)$ we may choose coordinates $(z,w) \in D \times D$ such that E_i has equation $w = 0$, where D is the unit disc in the complex number space. Let X be a holomorphic vector field on S tangent to the foliation. By means of π we obtain a holomorphic vector field X' on $D \times D^*$, where $D^* = D - \{0\}$. We are then reduced to showing that X' has at most a pole along $D \times 0$, since $D \times D$ is a contractible stein space and we may find a meromorphic function f whose zeroes and poles coincide with the poles and the codimension one subvarieties of $X' = 0$ respectively. The foliation may then be extends by the vector field fX on $D \times D$.

Since the tangent bundle to $D \times D^*$ is trivial, we have an expression for X' as a pair of functions $X' = (X'_1, X'_2)$ which may then be expanded in the form

$$X'_i = \sum_{m=-\infty}^{\infty} f^i_m(z) w^m$$

where f^i_m are holomorphic functions of z. X' has at most a pole along $D - 0$ if and only if $f^i_m = 0$ for $m < -N$. Note that for those $f^i_m(z)$ that are not identically equal to zero, at most a countable number of $z \in D$ satisfy at least one of the equations $f^i_m(z) = 0$, so that if X' had an essential singularity along $D=0$ we could find a $z \in D$ such that $X'(z_0,w)$ has an essential singularity at $w=0$.

We will now complete the proof by showing that for any $z \in D$, $X'(z,w)$ has at most a pole at $w=0$. Denote by C the curve $D \times 0$ and by Θ_S and $\Theta_{\tilde{S}}$ the coherent sheaves of holomorphic vector fields on S and $\tilde{S}$. The derivative of π defines a map $d\pi\colon \Theta_{\tilde{S}}|_C \to \pi^*\Theta_S|_C$ that is an isomorphism on $C-(0,0)$ and hece is injective. It may be completed to

$$0 \to \Theta_{\tilde{S}}|_C \to \pi^*\Theta_S|_C \to K \to 0 \tag{3}$$

where K is a coherent sheaf supported on $(0,0)$ and hece a skyscraper sheaf with a finite dimensional stem over $(0,0)$. Tensoring (3) with the sheaf of meromorphic functions M_C we obtain that $X \otimes M_C = 0$ since for big M, $Z^M K = 0$ by the finite dimensionality of K. Hence the sequence of Tor's of (3) gives in particular the surjective map

$$\Theta_{\tilde{S}}|_C \otimes M_C \to \pi^*\Theta_S|_C \otimes M_C \to 0$$

Hence we may find a meromorphic vector field Y on S restricted to C that is sent by $d\pi$ to $\pi^*(X)|_C$. By uniqueness it is clear that $Y(w) = X'(z_0,w)$ and hence X' has at most a pole along $D \times 0$ as needed. □

Remark: This Lemma is just a particular case of the behaviour of foliations under bimeromorphic maps.

The exact sequence of sheaves on $\tilde{S}$

$$0 \to Z \to \mathcal{O} \overset{e}{\to} \mathcal{O}^* \to 0 \tag{4}$$

induces the Chern map

$$C\colon H^1(\tilde{S},\mathcal{O}^*) \to H^2(\tilde{S},Z)$$

which associates to a line bundle its Chern class in $H^2(\tilde{S},Z)$. This last group is isomorphic to Z^n where n is the number of irreducible components E_i. If we denote by $[\cup E_i]$ the germ of $\tilde{S}$ around $\cup E_i$, the line bundles on $[\cup E_i]$ have been analyzed by Mumford [8].

THEOREM 12: *The exact sequence of sheaves (4) induces the exact sequence of groups*

$$0 \to \mathrm{Pic}_0[\cup E_i] \to H^1([\cup E_i],\mathcal{O}^*_{\tilde{S}}) \overset{c}{\to} H^2(\tilde{S},Z) \to 0$$

Where $H^1([\cup E_i],\mathcal{O}^*_{\tilde{S}})$ are the germs of holomorphic line bundles on $\tilde{S}$ around E_i, c is the Chern map and $\mathrm{Pic}_0[\cup E_i]$ defined by the kernel of c has a natural structure of a connected analytic group variety of the form $\mathbb{C}^N/\Lambda$, where Λ is a discrete lattice in $\mathbb{C}^N = H^1(\tilde{S},\mathcal{O}_{\tilde{S}})$.

We will analyse the foliation $\tilde{F}$ of Lemma 11 in a neighbourhood of $\cup E_i$. Since

S is smooth there is a line bundle $\tilde{\tau}$ defined on $\tilde{S}$ tangent to $\tilde{F}$. $\tilde{\tau}$ restricted to $S-UE_i$ is biholomorphic to τ on $S-\{0\}$, since there π sends $\tilde{F}$ to F biholomorphically. In this manner we may translate the problem of determining if F may be given by a vector field vanishing only at 0, via Proposition 10, to the problem of determining if $\tilde{\tau}$ restricted to $\tilde{S}-UE_i$ is analytically trivial near UE_i.

Let $\tilde{\varphi}: \tilde{\tau} \to T_{\tilde{S}}$ be the bundle map that determines the foliation $\tilde{F}$ on $\tilde{S}$. For each of the Riemann surfaces E_i there is a line bundle γ_i and a holomorphic section f_i of γ_i on $\tilde{S}$ that vanishes exactly on E_i.

Assume for a moment that $\tilde{\tau}$ is isomorphic to $\gamma_1^{m_1} \otimes \ldots \otimes \gamma_n^{m_n}$ as a holomorphic line bundle, and consider the map $\tilde{X} = \tilde{\varphi} f_1^{-m_1} \ldots f_n^{-m_n} : 1_{\tilde{S}} \to T_{\tilde{S}}$ obtained by multiplying $\tilde{\varphi}$ with $f_1^{-m_1} \ldots f_n^{-m_n}$. X has zeroes or poles along E_i, depending weather $m_i < 0$ or $m_i > 0$, but defined on the trivial line bundle $1_{\tilde{S}}$ by the assumptions on $\tilde{\tau}$. But then X may be viewed as a meromorphic vector field, that is non-vanishing on $S-UE_i$. Let X be the holomorphic vector field induced on $S' = S-\{0\}$ by means of $D\pi(X)$.

Then X is a holomorphic vector field on S' tangent to F that extends to a holomorphic vector field on S vanishing only at 0 (see the proof of theorem 5).

Conversely, assume that F is given by a vector field $\tilde{X}$ vanishing only at 0. Let $\tilde{X}$ be the meromorphic vector field obtained by π^{-1} on $\tilde{S}$, as in Lemma 11. The zeroes or poles of $\tilde{X}$ are concentrated on UE_i, so after cancelling with $f_i^{-m_i}$ we obtain that $f_1^{m_1}, \ldots, f_n^{m_n}\ \tilde{X}: \gamma_1^{m_1} \otimes \ldots \otimes \gamma_n^{m_n} \to T_{\tilde{S}}$ is a holomorphic map vanishing only at points, so $\tau = \gamma_1^{m_1} \otimes \ldots \otimes \gamma_n^{m_n}$. We have proved the preliminary

LEMMA 13: *F can be defined by a vector field vanishing at 0 if and only if $\tilde{\tau} = \gamma_1^{m_1} \otimes \ldots \otimes \gamma_n^{m_n}$ for some integers* $m_i \in Z$.

This lemma then will serve to provide topological and analytic obstructions, that is, when is the Chern class of $\tilde{\tau}$ of the form $\Sigma_{m_i} c(\gamma_i)$, and if it does, when is it analytically equivalent to $\gamma_1^{m_1} \otimes \ldots \otimes \gamma_n^{m_n}$?

We begin by analyzing the subgroup of $H^2(\tilde{S},Z)$ formed by $c(\gamma_1^{m_1} \otimes \ldots \otimes \gamma_n^{m_n})$. We form a basis of $H^2(S,Z)$ in the following form: choose analytic curves $\alpha_1,\ldots,\alpha_n$ in S transversal respectively to $E_1,\ldots,E_n$ such that $\alpha_i \cap E_i \notin E_j$ for $j \neq i$. The Chern classes of the line bundles associated to α_i form a basis of $H^2(S,Z) = Z^n$. Given a line bundle ξ on S, its Chern class in this basis is obtained by restricting ξ on E_i, and taking its Chern class on E_i, $c(\xi) \in H^2(E_i,Z) = Z$. In particular for γ_j, $c(\gamma_j) = (c(\gamma_j|E_1),\ldots,c(\gamma_j|E_n)) = (E_j \cdot E_1,\ldots,E_j \cdot E_n)$ where $E_j \cdot E_i$ is the intersection product in $H_2(\tilde{S},Z)$.

Consider the map

$$\Phi\colon Z^n \to H^2(S,Z) = Z^n$$

which associates to each $(m_1,\ldots m_n)$ the Chern class of the bundle $\gamma_1^{m_1} \otimes\ldots\otimes \gamma_n^{m_n}$ expressed in the above basis of $H^2(\tilde{S},Z)$. It is clear that Φ attains in matrix form the form of the intersection pairing of $H_2(\tilde{S},Z)$: $(E_i\cdot E_j)$. If we denote by $\mu = |\det\Phi|$ and by T the cokernel of Φ, then T is a finite group of order μ. Define

$$t\colon H^1([UE_i],\mathcal{O}^*) \to T$$

the map which associates first the Chern class of the bundle and then the projection to T. We can now phrase our result as

THEOREM 14: *Let* F *be a foliation on the normal surface* S, *and let* $\tilde{\tau}$ *be the line bundle tangent to the foliation* $\tilde{F}$ *induced on a desingularization* $\tilde{S}$ *of* S. F *may be defined by a holomorphic vector field in a neighbourhood of a singular point* p *if and only if*

1) *the topological obstruction vanishes*

$$t(\tilde{\tau}) = 0$$

2) *it the topological obstruction vanishes and the Chern class of* $\tilde{\tau}$ *is then equal to the Chern class of* $\gamma_1^{m_1} \otimes\ldots\otimes \gamma_n^{m_n}$ *then* $\tilde{\tau} \otimes \gamma_1^{-m_1} \otimes\ldots\otimes \gamma_n^{-m_n}$ *is the 0-element of the connected analytic group* $\mathrm{Pic}_0[UE_i]$ *of theorem 12.*

Proof: $\Rightarrow$) This is a direct application of the Lemma 12 and the definitions.

$\Leftarrow$) If 1) and 2) are satisfied, then $\tau = \gamma_1^{m_1} \otimes\ldots\otimes \gamma_n^{m_n}$ and again apply Lemma 13. □

COROLLARY 15: *Let* S *be a normal surface with a rational singularity. Then the foliation* F *is defined locally by a holomorphic vector field vanishing only at* 0 *if and only if* $t(\tilde{\tau}) = 0$.

Proof: By definition, a rational singularity is one where $H^1(\tilde{S},\mathcal{O}_{\tilde{S}}) = 0$ so that $\mathrm{Pic}_0([UE_i]) = 0$. Hence the analytic obstruction always cancels. □

There is a simple method to calculate the topological obstruction of theorem 14 that we now describe. Suppose X defines a foliation on the normal surface S with isolated singularity at 0, and let $\ell_1 C_1,\ldots,\ell_r C_r$ be the set of curves at 0 where X vanishes with multiplicities. Let $\tilde{X}$ be the meromorphic vector field induced on S by X, and assume the set of points where it vanishes or has poles has the form $\ell_1\tilde{C}_1 +\ldots+ \ell_r\tilde{C}_r + m_1E_1 + m_nE_n$ where $\tilde{C}_i$ is the proper transform of C_i, and m_i are some suitable integers. Thinking of the line bundle on $\tilde{S}\text{-}UE_i$, it has Chern class in $H^2(\tilde{S}\text{-}UE_i,Z) = H^2(K,Z) \cong H_1(K_1,Z)$ where K is the link of S and in the last equality we have used Poincaré-duality on the 3-manifold K. This element is easily seen to be $\ell_1(C_1\cap K) +\ldots+ \ell_r(C_r\cap K)$.

Mumford also shows that these elements always lie in the torsion part of $H_1(K,Z)$, which is a group with μ elements.

Summing up: Choose any holomorphic vector field X tangent to F on S. Let $(X)_0$ be the curves on which X vanishes, and consider the 1-cicle it determines in the link K of S, $(X)_0 \cap K \in H_1(K,Z)$. This element is the topological obstruction to defining F by a holomorphic vector field vanishing only at 0.

We now finish explaining Example 1. S is given by $Y^2 = XZ$ and it is easily seen that the desingularization of S is realized with one blow up of 0, and we obtain a rational curve with self intersection-2 (i.e. S is the cotangent bundle over $\mathbb{C}P^1$). In this case $T = Z/2Z = Z_2$.

Note also that the link $K = S^3/\underline{+}I$ by viewing $S = \mathbb{C}^2/\underline{+}I$. So by the above computations, if a foliation is given by vector field vanishing on $\ell_1C_1+\ldots+\ell_rC_r$, compute $\ell_1[C_1\cap K]+\ldots+\ell_r[C_r\cap K] \in H_1(K,Z) = Z_2$. This is the topological obstruction. By a direct calculation it is seen that S is a rational singularity, so this topological obstruction is the only obstruction.

Hence the dichotomy that appeared in Example 1, given by the algebraic condition $f^+(0) = \underline{+}\,1$ has then a topological interpretation: $f^+(0) = 1$ if the bundle is trivial and $f^+(0) = -1$ if it is not.

BIBLIOGRAPHY

[1] Camacho, C. *Quadratic Forms and the Separatrix Theorem for Singular Surfaces.* Preprint, IMPA, 1985.

[2] Fisher, G. *Complex Analytic Geometry.* LNM 538, Springer-Verlag, 1976.

[3] Gómez-Mont, X. *On Families of Rational Vector Fields.* Memorias del Coloquio de Sistemas Dinámicos, Guanajuato, México, 1985.

[4] Gómez-Mont, X. *Foliations by Curves of Complex Analytic Spaces.* Preprint, IMPA, 1985.

[5] Griffiths, PH., Harris, J. *Principles of Algebraic Geometry.* Wiley Interscience, 1978.

[6] Gunning, R., Rossi, H. *Analytic Functions of Several Complex Variables.* Prentice Hall, N.J., 1965.

[7] Hartshorne, R. *Algebraic Geometry.* Springer-Verlag, 1977.

[8] Mumford, D. *The Topology of Normal Singularities of an Algebraic Surface and a Criterion for Simplicity.* Publications IHES, No. 9, (1961), 5-22.

[9] Narasimhan, R. *Introduction to the Theory of Analytic Spaces.* LNM 25, Springer-Verlag, 1966.

[10] Rossi, H. *Vector Fields on Analytic Spaces.* Ann. Math. 78, (1963), 455-467.

INSTITUTO DE MATEMATICAS
Universidad Nacional Autónoma de México
México 04510, D.F.
MEXICO

Typist: Alma R. Fajardo M.

Contemporary Mathematics
Volume **58**, Part III, 1987

RENORMALIZATION OF ONE DIMENSIONAL MAPPINGS AND STRANGE ATTRACTORS

John Guckenheimer*

Jakobson [4] proved a remarkable theorem about the iteration of families of one dimensional mappings typified by the example $f_\mu(x) = \mu - x^2$, $\mu \in [1,2]$. For this example, his results state there is a set of μ having positive measure for which there is an absolutely continuous invariant ergodic measure for f_μ. The supports of these measures are "strange attractors" in the sense that they are the ω-limits of sets of trajectories having positive Lebesgue measure and they have a sensitive dependence to initial conditions. The proofs of Jakobson's Theorem and similar results of Collet and Eckmann [2] and Benedicks and Carleson [1] are complicated and involve intricate estimates. By using the idea of "renormalization" which has played an important role in understanding scaling properties of bifurcations, we extend portions of this work to give stronger results and simpler estimates. The thesis of Johnson [5] takes up questions concerning the existence of absolutely continuous invariant measures for the mappings studied in this paper. The results here concern topological properties of mappings associated with sets of parameter values having positive measure.

The conceptual basis for this paper lies with a recursion in which a geometric construction is followed by a rescaling of coordinates which reproduces a situation like that encountered at the beginning of the recursive step. Each step in the recursion requires that some parameter intervals from a family of mappings be excluded from further consideration because the geometric construction cannot proceed or because analytic estimates fail. However, at each step a smaller proportion of the remaining parameters are excluded, leaving a set of parameter values having positive measure for which the recursion can be carried forward without terminating.

To provide the setting for our construction, let us recall some topological properties of mappings of the interval with negative Schwarzian derivative.

---

* Research partially supported by National Science Foundation.

© 1987 American Mathematical Society
0271-4132/87 $1.00 + $.25 per page

Proposition [3]: If $f:I \to I$ has

(1) a single non-degenerate critical point c,

(2) $Sf(y) = \frac{D_{xxx}f(y)}{D_x f(y)} - \frac{3}{2}\left(\frac{D_{xx}f(y)}{D_x f(y)}\right)^2 < 0$ for all $y \in I-\{c\}$, and

(3) $f(\partial I) \subset \partial I$ with $|D_x f(y)| > 1$ for all $y \in \partial I$ a fixed point, then almost all trajectories of f have the same asymptotic behavior.

There are three possible types of asymptotic behavior allowed by this proposition:

(a) a stable periodic orbit

(b) A Cantor set Λ with $f|_\Lambda$ a 1-1 map

(c) a finite union of intervals on which f has dependence to initial conditions.

Case (c) of this list is the one which gives strange attractors. Some mappings have attractors which consist of a single interval $[f^2(c),f(c)]$. These mappings are characterized by the topological property that there is no proper subinterval $J \subset [f_2(c),f(c)]$ and integer $n > 0$ with $f^n(J) \subset J$. Our recursive construction will identify parameter values which yiedl mappings with this type of connected attractor. We shall not assume that our mappings have negative Schwarzian derivatie; all estimates involve only first and second derivatives.

Theorem: Let f_μ be a parameter family of mappings such that

(1) $f_\mu: [-1,1] \to [-1,1]$ with $f_u(\pm 1) = -1$ and $|D_x f_\mu(-1)| > 1$

(2) f_μ is a C^2 diffeomorphism of (x^2,μ) so that f_μ has a single nondegenerate critical point at 0

(3) there are parameter values μ_0 and μ_1 such that $f_{\mu_0}(0)=0$ and $f_{\mu_1}(0) = 1$. The mapping f_μ has no stable periodic orbits. Moreover $D_\mu f_\mu(0) > 0$.

Then the set of μ such that f_μ has a strange attractor has positive Lebesgue measure.

The proof of this theorem occupies the next three sections of this paper.

1. *The Renormalized Induced Map*

In this section we describe the geometric construction which underlies the recursion employed in our proof of the Theorem. The following notation will be employed throughout the paper: $I = [-1,1]$, the Lebesgue measure of a set E is denoted $\ell(E)$, derivatives of a family of functions $f_\mu: I \to I$ are denoted $D_x f$, $D_\mu f$, $D_{x\mu} f$, etc., the closure of a set E is denoted $\bar{E}$, and the boundary of an interval J is denoted ∂J. If f can be written in the form $f(x)=g(x^2)$ with g r times differentiable, then we say that f is *strongly* C^r.

Let $f: I \to I$ be a mapping with the following properties:

(11) There are disjoint intervals I_j, $J \subset I$ such that $I - (\cup I_j \cup J)$ has Lebesgue measure zero

(12) $f: I_j \to I$ extends to a C^2 monotone mapping of I_j *onto* I.

(13) $0 \in J$ and f extends to a strongly C^2 mapping of $\bar{J}$ onto I with a single critical point at 0 and $f(\partial\bar{J}) \subset \partial I$.

We call J the *central* interval of I. The *induced mapping* $g: \bar{J} \to \bar{J}$ is defined by $g(x) = f^n(x)$ with n the smallest positive integer such that $f^n(x) \in \bar{J}$. If $f^n(x) \notin J$ for all $n > 0$, then $g(x)$ is not defined. We use the convention that if f is defined on $I_j = (a_-,a_+)$, then we write $f(a_\pm) = \lim_{x \to a_\pm^\mp} f(x)$ even though this may be ambiguous because $\bar{I}_j \cap \bar{I}_k \neq \emptyset$.

We want to carry through the construction of induced mappings of induced mappings of induced mappings, etc. In a family of mappings, we seek parameter values for which the construction can be carried out recursively in such a way that properties (11)-(13) are always satisfied with the additional requirement that $f(0)$ lies outside the central interval. These mappings will be ones whose limit sets are strange attractors. The initial step of the recursion is different from the remainder because we start with a smooth mapping $f: I \to I$.

LEMMA 1.1 If $f: I \to I$ satisfies properties (11)-(13) and if $\{x \in J | f^n(x) \in J$ for all $n > 0\}$ has measure zero, then the induced mapping $g: J \to J$ satisfied (11)-(13) (with central interval $J' \subset J$ and I replaced by J).

Proof: The intervals J' and $J_i \subset J$ are defined to be the components of the domain of g. We have assumed that g is defined almost everywhere, so (11) is satisfied. The chain rule implies that 0 is the only critical point of g. Properties (12) and (13) could fail only if there is an endpoint of J_i of J' with $g(a)$ in the interior of J(with $g(a)$ defined as a one-sided limit as above). If $g(a) = f^n(a)$, the $f^i(z) \in I_{j_i}$ for $i < n$ because points not in the I_{j_i} of J do not return to J. We conclude that f^n is defined and continous in a neighborhood of a, implying that a is in the interior of the domain of g and contradicting that it was and endpoint of a component of the domain of g.

Determining whether and induced mapping satisfied (11) involves two issues:

the possible existence of stable periodic orbits which lie entirely outside J and, in the absence of stable periodic orbits, the measure of the nowhere dense set of trajectories which remain outside of J. Stable periodic orbits outside J can be precluded by the hypothesis that $|D_x f| > 1$ in UI_j. The second issue can be resolved with estimates of the size of $|D_{xx}f/(D_x f)^2|$ which enter the recursion defined in Section 2. To obtain an initial mapping satisfying (11)-(13), select and unstable periodic point p in the interior of I with the property that if $f(p') = f(p)$, then no point of the orbit p lies in (p,p'). If the set of trajectories remaining outside (p,p') has measure zero, then we can begin the recursion with the induced mapping for $J = (p,p')$. In particular, Guckenheimer [3] implies that this is true for the family f_μ of the theorem when p is the fixed point of f interior to I and μ is near μ_1.

Before proceeding to the next section and its analytic estimates, we introduce a simple example which motivates the strategy employed in the recursion. Call an <u>LQ</u> (linear-quadratic) mapping $f\colon I \to I$ one which satisfies (11)-(13) with $f|_{I_j}$ linear and $f|_J$ quadratic. Consider a family $f_\mu\colon I \to I$ of LQ maps in which $f_\mu(x) = \mu - \frac{S}{2}x^2$ for all $x \in J_\mu$ with S independent of μ and $\mu \in [-1,1]$. For this LQ family, we want to estimate the measure of parameter values to be excluded from further consideration because there is a proper subinterval mapped into itself by f_μ or by g_μ. By solving the equation $\mu - \frac{S}{2}x^2 = -1$, we compute $\ell(J_\mu) = 2(2(1+\mu)/S)^{1/2}$. If S is large and $(\mu+1)$ is not small relative to $S^{-1/2}$, then $\ell(J_\mu)$ is of order $S^{-1/2}$. The set of parameter values to be excluded because there is a subinterval $J' \subset J$ with $f_\mu(J') \subset J'$ thus has lenght of order $S^{-1/2}$.

The induced mapping g_μ of f_μ is again a quadratic function on its central interval because it is a composition of several functions with one quadratic function. The size of its central interval J' will determine the size of the parameter interval to be excluded because g_μ maps a subinterval of J' into itself. In a parameter range for wich g_μ is continuously given by f_μ^n, we want a significantly smaller proportion than $S^{-1/2}$ to be excluded. If $|D_\mu g_m(0)|$ is roughly constant and $|D_{xx}g(0)| >> S^2$, then $\ell(J') - |D_{xx}g(0)|^{-1/2} << S^{-1}$ and $\ell(J')/\ell(J) << \ell(J)/\ell(I)$. If $|D_x f| > L$ on UI_j, then the estimate $|D_{xx}g(0)| >> S^2$ requires that $L^n >> S$. For example, we might require $n > 2\log S/\log L$. In parameter intervals for which the return time of 0 is at least this long, the proportion of parameters excluded because g_μ has an invariant subinterval is significantly smaller than the proportion of the original parameter interval excluded because f_μ had an invariant subinterval.

This motivates us to eliminate parameter intervals with "short" return times provided that the total length of these is small. Define $E_{\mu,n} = E_{\mu,n-1} \cap f^{-n}(I-J)$ with $E_{\mu,o} = I-J$. Inductively, one sees that each component of $E_{\mu,n}$ is mapped linearly onto I. Therefore a proportion $\ell(J)/\ell(I)$ of the points in $E_{\mu,n-1}$ do not lie in $E_{\mu,n}$. We conclude that $\ell(E_{\mu,n}) = \ell(I-J)(\ell(I-J)/\ell(I))^n$. With $n \sim \log S$ and $\ell(I-J)/\ell(I) \sim 1-S^{-1/2}$, we have $\ell(E_{\mu,n})/2 \sim (1 - S^{-1/2})^{\log S}$. As $S \to \infty, (1-S^{-1/2})^{\log S} \to 1$, so as S increases, all but a small proportion of the parameter values give long enough return times for 0 to guarantee the estimate $|D_{xx}g(0)| > S^3$. This is a key observation which underlies our proof of the Theorem. At each step of the recursion we exclude parameter values giving short return times as well as parameter values for which we lose control of the distortion of our mappings from an LQ example. The distortion estimates are the subject of Section 2 and the parameter exclusions are in Section 3.

2. *Recursive Hypothesis and Estimates*

This section formulates the analytic estimates which enter into the recursion for the proof of the Theorem. These estimates are used in Section 3. Throughout the remainder of the paper, we assume that $f_\mu: I \to I$, $I = [-1,1]$, $\mu \in [0,1]$, is a family of mappings satisfying the following hypotheses:

(RHO) $I = \cup I_{j,\mu} \cup J_\mu$ (mod sets of measure zero) and $f_\mu|I_{j,\mu}$, $f_\mu|J_\mu$ are strongly C^2 mappings. The mapping f_μ is even: $f_\mu(-x) = f_\mu(x)$.

(RHI) $f_\mu|I_j$ maps I_j onto I with

(a) $|D_x f| \geq L > 2$

(b) $\left|\dfrac{D_{xx}f}{(D_x f)^2}\right| \leq D_1$

(c) $\left|\dfrac{D_{\mu x}f}{(D_x f)^2}\right| \leq D_3$

(d) $\left|\dfrac{D_{\mu\mu}f}{(D_x f)^2}\right| \leq D_5$

(RH2) $f_\mu | J_\mu$ has a single critical point at 0, $f_\mu(\partial J_\mu) \subset \partial I$, and

(a) $\left|\dfrac{D_{xx}f(y)}{D_x f(y)} - \dfrac{1}{y}\right| \left|\dfrac{1}{D_x f(y)}\right| \le D_2$ for all $y \in J - \{0\}$.

(b) $\left|\dfrac{D_{\mu x}f}{D_\mu f D_x f}\right| \le D_4$

(c) $\left|\dfrac{D_{\mu\mu}f}{(D_\mu f)^2}\right| \le D_6$

(RH3) There are constants λ and $E < \frac{L-1}{2L}$ such that

(a) $f_\mu | I_{j,\mu}$ satisfies $\left|\dfrac{D_\mu f}{D_x f}\right| < \lambda E$

(b) $f_\mu | J_\mu$ satisfies $|D_\mu f| > \lambda$

The following Lemmas (2.1)-(2.4) give estimates corresponding to hypotheses (RH1) and (RH2) for the induced mapping g. We continue to use the notation $E_{\mu,n} = \{x \in I | f_\mu^i(x) \notin J_\mu$ for $0 \le i < n\}$. We also write $g(x) = h\, f(x)$ where $h = f^n$ if $f(x) \in E_{\mu,n} - E_{\mu,n+1}$.

Lemma 2.1 If $y \in E_{\mu,n}$, then $\left|\dfrac{D_{xx}f^n(y)}{D_x f^n(y))^2}\right| < \dfrac{D_1 L}{L-1}$. If $[y,z] \subset E_{\mu,n}$, then

$$e^{(-D_I\frac{L}{L-I}\,|\,f^n(z)-f^n(y)\,|\,)} < \frac{D_x f^n(z)}{D_x f^n(y)} < e^{(D_I\frac{L}{L-1}\,|\,f^n(z)-f^n(y)\,|\,)}.$$

Proof: Assume inductively that, for $k < n$,

$$\left|\frac{D_{xx}f^k(y)}{D_x f^k(y))^2}\right| < \frac{D_1 L}{L-1}$$

Compute

$$\left|\frac{D_{xx}f^{k+1}(y)}{(D_x f^{k+1}(y))^2}\right| \le \left|\frac{D_{xx}f(f^k(y))}{(D_x f(f^k(y)))^2}\right| + \left|\frac{1}{D_x f(f^k(y))}\right| \left|\frac{D_{xx}f^k(y)}{(D_x f^k(y))^2}\right|$$

$$< D_1 + \frac{1}{L}\left(D_I\frac{L}{L-1}\right) = D_I\,\frac{L}{L-1}$$

To prove the second inequality, we note that f^n is monotone on $[y,z]$ and

$$\left|\log \frac{D_x f^n(z)}{D_x f^n(y)}\right| = \left|\int_y^z \frac{D_{xx} f^n(\xi)}{D_x f^n(\xi)} d\xi\right| \leq \int_y^z D_1 \frac{L}{L-1} |D_x f^n(\xi)| \, d\xi$$

Lemma 2.2 If $y \in J-\{0\}$ and $g(y)$ is defined, then

$$\left|\frac{D_{xx} g(y)}{(D_x g(y))^2}\right| \leq D_1 \frac{L}{L-1} + \frac{1}{|D_x h(f(y))|}\left(D_2 + \frac{1}{|y||D_x f(y)|}\right)$$

Proof: Apply the chain rule to $g = h\,f$ to obtain

$$\left|\frac{D_{xx} g(y)}{(D_x g(y))^2}\right| \leq \left|\frac{D_{xx} hf(y))}{(D_x h(f(y)))^2}\right| + \frac{1}{|D_x h(f(y))|} \left|\frac{D_{xx} f(y)}{(D_x (f(y))^2}\right|$$

Now use Lemma (2.1) and (RH2) to estimate the magnitude of the two terms.

Lemma 2.3 If $y \in E_{\mu,n}$, then

(a) $$\left|\frac{D_\mu f^n(y)}{D_x f^n(y)}\right| < \lambda \frac{L}{L-1} E$$

(b) $$\left|\frac{D_{\mu x} f^n(y)}{(D_x f^n(y))^2}\right| \leq D_1 \lambda \left(\frac{L}{L-1}\right)^2 E + \frac{nD_3}{L^{n-1}}$$

(c) $$\left|\frac{D_{\mu\mu} f^n(y)}{(D_x f^n(y))^2}\right| \leq D_1 \lambda^2 \left(\frac{L}{L-1}\right)^3 E^2 + \frac{n^2 D_3}{L^{n-1}} (\lambda E) + \frac{D_5}{L^n (L-1)}$$

Proof: For (a), we compute

$$\frac{D_\mu (f^n \circ f)}{D_x (f^n \circ f)} = \frac{[(D_x f^n) \circ f] D_\mu f}{[(D_x f^n) \circ f] D_x f} + \frac{(D_\mu f^n) \circ f}{[(D_x f^n) \circ f]} \cdot \frac{1}{D_x f}$$

Since $|D_x f| > L$ by (RH1a) and $\left|\frac{D_\mu f}{D_x f}\right| < \lambda E$ by (RH3a),

$$\left|\frac{D_\mu(f^n \circ f)}{D_x(f^n \circ f)}\right| \leq \lambda E + \frac{\lambda EL}{L-1} \cdot \frac{1}{L} = \lambda E \frac{L}{L-1}$$

For (b),

$$\frac{D_{\mu x}(f^n\, f)}{(D_x(f^n\, f))^2} = \frac{D_{xx}f^n}{(D_xf^n)^2}\,\frac{D_\mu f}{D_xf} + \frac{1}{D_xf^n}\,\frac{D_{\mu x}f}{(D_xf)^2} + \frac{D_{\mu x}f^n}{(D_xf^n)^2}\,\frac{1}{D_xf}$$

If we assume inductively that (b) is satisfied for n then

$$\left|\frac{D_{\mu x}(f^n\, f)}{(D_x(f^n\, f))^2}\right| < D_1\frac{L}{L-1}\lambda E + \frac{1}{L^n}D_3 + (D_1\lambda(\frac{L}{L-1})^2E + \frac{nD_3}{L^{n-1}})\frac{1}{L}$$

$$= D_1(\frac{L}{L-1})^2\lambda E + \frac{n+1}{L^n}D_3 .$$

For (c)

$$\frac{D_{\mu\mu}f^n \circ f}{(D_xf^n \circ f)^2} = \frac{D_{xx}f^n}{(D_xf^n)^2}(\frac{D_\mu f}{D_xf})^2 + 2\frac{D_{\mu x}f^n}{(D_xf^n)^2}\frac{D_\mu f}{(D_xf)^2} + \frac{1}{D_xf^n}\frac{D_{\mu\mu}f}{(D_xf)^2} + \frac{D_{\mu\mu}f^n}{(D_xf^n)^2}\frac{1}{(D_xf)^2}$$

Assuming (c) for n, this yields

$$\left|\frac{D_{\mu\mu}f^n \circ f}{(D_xf^n \circ f)^2}\right| \leq D_1\frac{L}{L-1}(\lambda E)^2 + \frac{2}{L}[D_1\lambda(\frac{L}{L-1})^2E + \frac{nD_3}{L^{n-1}}](\lambda E) + \frac{1}{L^n}D_5$$

$$+ \frac{1}{L^2}[D_1\lambda^2(\frac{L}{L-1})^3E^2 + \frac{n^2D_3}{L^{n-1}}(\lambda E) + \frac{D_5}{L^n(L-1)}]$$

$$= D_1\frac{L}{L-1}(\lambda E)^2[1 + \frac{2}{L-1} + \frac{1}{(L-1)^2}] + D_3\frac{\lambda E}{L^n}(2n + \frac{n^2}{L}) + \frac{D_5}{L^n}(1 + \frac{1}{L^2(L-1)}).$$

$$< D_1(\frac{L}{L-1})^3(\lambda E)^2 + D_3\frac{\lambda E}{L^n}(n+1)^2 + \frac{D_5}{L^n(L-1)}$$

<u>Lemma 2.4</u> If $y \in J-\{0\}$ and $g(y) = f^{n+1}(y)$, then

(a) $|D_\mu g(y)| > \frac{\lambda L^n}{2}$

(b) $\left|\frac{D_{\mu x} g(y)}{D_\mu g(y) D_x g(y)}\right| \leq 3D\frac{L}{IL-1} + \frac{2nD_3}{\lambda L^{n-1}} + \frac{2D_4}{L^n}$

(c) $\left|\frac{D_{\mu\mu} g(y)}{(D_\mu g(y))^2}\right| \leq 9D\frac{L}{IL-1} + \frac{4(n+1)^2 D_3}{\lambda L^{n-1}} + \frac{4D_5}{\lambda^2 L^n (L-1)} + \frac{4D_6}{L^n}$

Proof: Write $g = h \quad f$, so that

$$D_\mu g = D_x h(D_\mu f + \frac{D_\mu h}{D_x h})$$

Since $E < \frac{L-1}{2L}$, (RH3b) and Lemma (2.3a) imply that $|D_\mu g(y)| < |D_x h(f(y))| \cdot \frac{\lambda}{2}$. This yields (a) because $|D_x h| \geq L^n$. For (b), we have $|D_\mu f + \frac{D_\mu h}{D_x h}| \leq \frac{1}{2} |D_\mu f|$ which yields

$$\frac{D_{\mu x} g}{|D_\mu g D_x g|} \leq 2[\left|\frac{D_{xx} h}{(D_x h)^2}\right| + \frac{1}{|D_\mu f|}\left|\frac{D_{x\mu} h}{(D_x h)^2}\right| + \frac{1}{|D_x h|}\left|\frac{D_{x\mu} f}{D_x f D_\mu f}\right|]$$

$$\leq 2[D\frac{L}{1L-1}(1 + \frac{L}{L-1}E) + \frac{nD_3}{\lambda L^{n-1}} + \frac{D_4}{L^n}]$$

Since $\frac{L}{L-1}E < \frac{1}{2}$, this gives (b). For (c) we estimate

$$\frac{D_{\mu\mu} g}{|(D_\mu g)^2|} \leq 4[\left|\frac{D_{xx} h}{(D_x h)^2}\right| + \frac{2}{|D_\mu f|}\left|\frac{D_{x\mu} h}{(D_x h)^2}\right| + \frac{1}{|D_x h|}\left|\frac{D_{\mu\mu} f}{(D_\mu f)^2}\right| + \left|\frac{D_{\mu\mu} h}{(D_x h)^2}\right|]$$

$$\leq 4[D\frac{L}{IL-1}(1 + \frac{2L}{L-1}E + (\frac{LE}{L-1})^2) + \frac{D_3}{\lambda L^{n-1}}(2n + n^2 E) + \frac{D_5}{\lambda^2 L^n (L-1)} + \frac{D_6}{L^n}]$$

In proceeding with the recursion, we want to retain parameter values for which the hypotheses (RH) are satisfied. These are not invariant under linear corrdinate changes, so it is helpful to rescale g so that the lengths of J and the interval of parameter variation are restored to 2 and 1, respectively. If $[\mu_1, \mu_2]$ is a maximal parameter interval for which the exponent in $g_\mu(0) = f_\mu^{n-1}(0)$ remains constant, then we define the *renormalized induced mapping* by $\bar{g}_\nu(y)=\alpha^{1}g_{(\beta\nu+\mu_1)}(\alpha Y)$ where $\alpha = \ell(J_\mu)/2$ and $\beta = (\mu_2-\mu_1)$. Thus $\bar{g}$ is defined on the interval $[-1,1]$ for all parameter values and ν varies in the interval $[0,1]$. In the recursion, we want $\bar{g}$ to satisfy the (RH) hypotheses with new values $(\bar{D}_i,\bar{\lambda},\bar{E})$ of the constants (D_i,λ,E) which will be specified as part of the recursion. The relationship between the derivatives of $\bar{g}$ and g is given by

$$\frac{\partial^{i+j}}{\partial y^i \partial \nu^i}(\bar{g}) = \alpha^{(i-1)}\beta^j \frac{\partial^{(i+j)}}{\partial x^i \partial \mu^j}(g) .$$

This formula, together with the estimates of Lemmas (2.1)-(2.4) gives the following recursive bounds for $(\bar{D}_i,\bar{E})$:

$$\text{(RB1)}\ \bar{D}_1 = \alpha[D_1\frac{L}{L-1} + \sup_{y\in J-J'}(\frac{1}{|D_xh(f(y))|}(D_2 + \frac{1}{|y||D_xf(y)|}))]$$

$$\text{(RB2)}\ \bar{D}_3 = \beta[3D_1\frac{1}{L-1} + \frac{2nD_3}{\lambda L^{n-1}} + \frac{2D_4}{L^n}]\sup_{y\in J-J'}\left|\frac{D_\mu g(y)}{D_x g(y)}\right|$$

$$\text{(RB3)}\ \bar{D}_5 = \beta^2[9D_1(\frac{L}{L-1} + \frac{4((n+1)^2D_3}{\lambda L^{n-1}} + \frac{4D_5}{\lambda^2L^n(L-1)} + \frac{4D_6}{L^n}](\sup_{y\in J-j'}\left|\frac{D_\mu g(y)}{D_x g(y)}\right|)^2$$

$$\text{(RB4)}\ \bar{D}_2 = \alpha[D_1(\frac{L}{L-1}) + \frac{D_2}{\inf_{y\in J'}|D_xh(f(y))|}]$$

$$\text{(RB5)}\ \bar{D}_4 = \alpha[3D_1(\frac{L}{L-1} + \frac{2nD_3}{\lambda L^{n-1}} + \frac{2D_4}{L^n}]$$

$$\text{(RB6)}\ \bar{D}_6 = \alpha[9D_1(\frac{L}{L-1}) + \frac{4(n+1)^2D_3}{\lambda L^{n-1}} + \frac{4D_5}{\lambda^2L^n(L-1)} + \frac{4D_6}{L^n}]$$

$$\text{(RB6)}\ \bar{E} = (\sup_{\substack{y\in J-J'\\ z\in J'}}\frac{1}{|D_xf(y)||D_xh(f(z))|})(\frac{1 + E\frac{L}{L-1}}{1 - E\frac{L}{L-1}})$$

3. *Parameter Exclusions*

Lemma 3.1 Let $f: I \to I$ satisfy (RH0), (RH1a) and (RH1b). If Λ is the set of points whose trajectories remain outside J, and $\ell(J) < 2e^{2D_1\frac{L}{L-1}}$, then Λ has measure zero.

Proof: Recall that $E_{\mu,n} = \{x|f^i(x) \notin J \text{ for } 0 \leq 1 < n\}$. Consider a component K of $E_{\mu,n}$ and $f^n|_K$ which maps K monotonically onto I. We estimate $\ell(K \cap E_{\mu,n+1})/\ell(K)$ by using Lemma 2.1 and the mean value theorem. There are points $y,z \in K$ such that $|D_x f^{n+1}(y)|\ell(K\cap f^{-(n+1)}(J)) = \ell(J)$ and $|D_x f^{n+1}(z)|\ell(K)=\ell(I)=2$. Therefore $\ell(K\cap E_{\mu,n+1})/\ell(K) < (1-\frac{\ell(J)}{2}e^{-2D_1\frac{L}{L-1}}) < 1$. Thus, we conclude inductively that

$$\ell(E_{\mu,n} < (2 - \ell(J))(1 - \frac{\ell(J)}{2}e^{-2D_I\frac{L}{L-1}})^{n-1}$$

and this tends to zero as $n \to \infty$.

We obtain a lower bound on $\ell(E_{\mu,n})$ from this argument by using the other inequality for $|Df^{n+1}(y)|/|Df^{n+1}(z)|$ from Lemma 2.1.

Lemma 3.2. $\ell(E_{\mu,n}) \geq (2 - \ell(J))(1 \frac{\ell(J)}{2}e^{2D_I\frac{L}{L-1}})^{n-1}$.

Proof: As above.

Observe that for fixed D_1, $\ell(E_{\mu,n})/\ell(I) \to$ as $\ell(J) \to 0$. Lemma 3.2 gives a quantitative estimate for the expected length of return times for a large sets of the critical point for $f_\mu(0)$ moves across the sets Λ_μ. The parameter exclusions formulated below specify that $f_\mu(0)$ be situated properly relative to Λ_μ. The exclusions are necessary to obtain estimates for the size of the recursive bounds in (RB). The proof of the theorem ends with a description of how $f_\mu(0)$ moves across Λ_μ as μ is varied. There are four types of parameter exclusioms. The first guarantees that return times are long, while the final three guarantee that $f_\mu(0)$ is relatively far from endpoints $E_{\mu,n}$.

(PEI) $f_\mu(0) \in E_{\mu,n}$ for some $n > \left[-\frac{3\log|D_{xx}f(0)|}{\log L}\right]$

Lemma 3.3: $\ell(J) \leq 4e^{D_2}|D_{xx}f(0)|^{-1/2}$.

Proof: If $y \in J$, then (RH2a) implies

$$\left|\log \frac{D_{xx}f(0)}{\frac{D_x f(y)}{y}}\right| = \left|\int_0^y \left(\frac{D_{xx}f(\xi)}{D_x f(\xi)} - \frac{1}{\xi}\right) |d\xi|\right| \leq D_2 \left|\int_0^y |D_x f(\xi)|\,|d\xi|\right| \leq 2D_2$$

Therefore, $|y| \leq |D_x f(y)|\,|D_{xx}f(0)|^{-1} 2e^{2D_2}$. Letting y vary from 0 to an endpoint z of J, we obtain

$$\frac{z^2}{2} \leq \ell(f(J))|D_{xx}f(0)|^{-1}e^{2D_2} \leq 2|D_{xx}f(0)|^{-1}e^{2D_2}$$

Thus

$$\frac{\ell(J)}{2} \leq |z| \leq 2|D_{xx}f(0)|^{-1/2}e^{D_2}$$

Denote $D_{xx}f(0)$ by S for the remainder of this section.

Corollary 3.4: The set of values for $f_\mu(0)$ allowed by PE(1) is at least

$$\ell(E_{\mu,n_0}) \geq (2 - 4S^{1/2}e^{D_2})(1 - 2S^{1/2}e^{(D_2 + 2D_1\frac{L}{L-1})})^{\frac{3\log S}{\log L}}$$

where n_0 is the smallest integer larger than $\frac{3\log S}{\log L}$.

The next two parameter exclusions are concerned with the recursive hypothesis (RH1a) and (RH1b) for the induced mapping. If $y \in J - J'$, then we want and estimate for the return time of y in terms of $|y|$. Small values of $|y|$ require long return times so that $|D_x g(y)|$ will be larger than L and satisfy the distortion estimates of (RH1b). We motivate these exclusions as follows. If Γ is a component of the complement of Λ with $f^n(\Gamma) = J$, and if $f(y) \in \Gamma$ but $f(0) \notin \Gamma$, then $f(0)$ cannot lie "too close" to Γ. In terms of $\ell(\Gamma)$, $D_x g = D_x h D_x f$ and $D_x h$ is of order $\ell(J)/\ell(\Gamma)$. Thus we need to exclude a range of values for $f_\mu(0)$ outside Γ whose length is of order $LS^{-1}\ell(\Gamma)/\ell(J)$.

Precisely, points $u, v \in \Gamma$ satisfy

$$\left|\frac{D_x h(u)}{D_x h(v)}\right| < e^{D\frac{L}{IL-1}\ell(J)}$$

Therefore, since $\frac{L}{L-1} < 2$,

$$|D_x(h)| > \frac{\ell(J)}{\ell(\Gamma)} e^{-2D_1\ell(J)} \text{ for al } u \in \Gamma$$

Moreover, RH(2a) implies for $z \in |0,y|$ that

$$S|z|e^{-D_2|f(z)-f(0)|} \le |D_x f(z)| \le S|z|e^{D_2|f(z)-f(0)|}$$

Hence

$$|f(y) - f(0)| \le \frac{y^2}{2} Se^{D_2|f(y)-f(0)|}$$

and

$$\sqrt{2}|f(y) - f(0)|^{1/2}S^{1/2}e^{-(3/2)D_2|f(y)-f(0)|} \le |D_x f(y)|$$

Therefore, if $f(y) \in \Gamma$ and $f(0) \notin \Gamma$, we require that

---

$$\text{(PE2)} \quad |f(y) - f(0)|e^{-3D_2|f(y)-f(0)|} \ge (\ell(\Gamma))^2 \left| \frac{L^2}{2S(\ell(J))^2} \right| e^{4D_1\ell(J)}$$

When (PE2) is satisfied for all y, then the renormalized induced mapping satisfies (RH1a).

To obtain the recursive estimate for (RH1b) we examine (RB1). Using the estimates for $D_x f(y)$ and $D_x h(f(y))$ obtained above in (RH1b), we required a bound on

$$\frac{\ell(J)}{2} \frac{1}{|D_x h(f(y))|} \frac{1}{|y||D_x f(y)|} \le \frac{\ell(J)}{2} \frac{\ell(\Gamma)}{\ell(J)} \left(e^{2D_1\ell(J)}\right)\left(\frac{1}{Sy^2}\right)e^{D_2|f(y)-f(0)|}\,)$$

$$\le \ell(\Gamma)e^{2D_1\ell(J)+2D_2|f(y)-f(0)|} \cdot \frac{1}{|f(y)-f(0)|}$$

in terms of $\bar{D}_1$. (The last inequality is obtained by integrating $|D_x f(z)| \le S|z|e^{D_2|f(y)-f(0)|}$ over the interval $|0,y|$ to estimate $\frac{1}{Sy^2}$. If D_1 were to remain bounded throughout the recursion, then the lengths of the excluded parameter intervals would be proportional to $\ell(\Gamma)$ and we would not have proportionately smaller exclusions as the recursion proceeds. Therefore, we shallow allow D_1 to grow by setting $\bar{D}_1 = 2D_1$. We then require that if $f(y) \in \Gamma$ and $f(0) \notin \Gamma$ that

$$\text{(PE3)}\quad |f(y) - f(0)|e^{-2D_2|f(y)-f(0)|} \geq \frac{\ell(\Gamma)}{(2-\ell(J))D_1-D_2} e^{2D_1\ell(J)}$$

When (PE3) is satisfied, the renormalized induced mapping satisfies (RH1b) with D_1 replaced by $\bar{D}_1 = 2D_1$.

The final parameter exclusion involves the bound (RB7). In (RB7), let u be the endpoint of J' with the same sign as z. Then

$$\left|\frac{D_x f(u)}{D_x f(y)}\right| \leq \left|\frac{u}{y}\right| e^{D_2|f(y)-f(u)|} \; e^{2D_2}$$

$$\left|\frac{D_x h(f(u))}{D_x h(f(z))}\right| \leq e^{D\frac{L}{IL-I}\ell(J)} \leq e^{2D_1\ell(J)}$$

Therefore we can choose $\bar{E}$ to be any number such that

$$\bar{E} \geq \frac{1}{|D_x g(u)|} e^{(2D_1\ell(J)+2D_2)} \frac{1+2E}{(1-2E)}$$

Since E is to be small, we want to exclude parametre values for which $|D_x g(u)|$ is small. Assume that $f_\mu(0)$ lies in a component Γ of the complement of Λ. We then have the estimate

$$|D_x g(u)| \geq \frac{\ell(J)}{\ell(\Gamma)} e^{-D_1\ell(J)})(\sqrt{2}\, \bar{s}^{1/2}|f(u) - f(0)|^{1/2} e^{-\frac{3}{2}D_2|f(u) - f(0)|})$$

Thus if $f(0) \in \Gamma$, u is the endpoint of Γ in $f(J)$, and $\bar{E}$ is specified, we require

---

$$(PE4)\quad |f(u) - f(0)|e^{-3D_2|f(u)-f(0)|} \geq \frac{1}{2\overline{E}\ {}^2S}\ \frac{\ell(\Gamma)}{(\ell(J))^2}\ \frac{1+2E}{(1-2E)^2}\ 6D_1\ell(J)+4D_2$$

It suffices to take $E = \overline{E} = \frac{1}{4}$ in the proof of the Theorem.

The next step in proving the Theorem is an examination of how the measure of the set of parameter exclusions depends upon the constants D_i, E, L, S. Initially we are interested in the limiting behavior as $S \to \infty$ with D_1, D_2, E and L fixed.

Lemma 3.5: If D_1, D_2, L and E are fixed with $2D_1 - D_2 > 0$ and if $\epsilon > 0$, there is an $S > 0$ such that the set of values for f(0) which fail any of the conditions (PE1),(PE3) or (PE4) has measure smaller than ϵ.

Proof: Corollary 3.4 implies that the set of values which fail (PE1) has measure tending to 0 because $(1 - 2S^{-1/2}e^{(D_2+4D_1)})^{\frac{3\log S}{\log L}} \to 1$ as $S \to \infty$. With (PE3) and (PE4), the range of values of f(0) which is excluded due to the component Γ of $I - \Lambda$ has length which is proportional to $\ell(\Gamma) \cdot [\frac{\ell(\Gamma)}{S(\ell(J))}2]$. From the estimate used in formulating (PE2), we know that

$$2\ell(f(J)) \leq \frac{\ell(J)}{(\ \ 2)}{}^2Se^{D_2\ell(f(J))}$$

or

$$8\ell(f(J))e^{D_2\ell(f(J))} \leq \ell(J)^2S.$$

Thus the excluded ranges for f(0) have length of order $\frac{\ell(\Gamma)^2}{\ell(f(J))}$. Since $\ell(\Gamma) < \ell(J) < 4e^{D_2}S^{-1/4}$ provided that $\ell(f(J)) > S^{-1/4}$. Ignoring the small range for f(0) which $1 - |f(0)| \leq S^{-1/4}$, we see that ranges of f(0) excluded by (PE3) and (PE4) have total length which tends to 0 as $S \to \infty$.

The exclusions from (PE2) require that we let D_1 increase during the recursion, as noted above. This motivates the following proposition.

Proposition: Given positive numbers D_1, $\overline{D}_1 (1 \leq i \leq 6)$, E and $\overline{E}$, there is a number $S_0 > 2$ such that if $S \geq S_0$ and if f_μ satisfies the recursive hypotheses (RH) and the parameter exclusions (PE), then the renormalized induced mapping $\overline{g}_\mu$ satisfies the recursive hypotheses with constants $\overline{D}_i$ replacing D_i and $\overline{E}$ replacing E. Moreover, $S_0(2D_1,\ldots,D_6,E;\ 4D_1,D_2,\ldots,D_6,E) < S_0^3(D_1,\ldots,D_6,E;\ 2D_1,D_2,\ldots,D_6,E)$.

Proof: This is a matter of checking that the parameter exclusions and S large imply that the recursive bounds (RB) are satisfied. The estimates (RB1) and (RB7) are consequences of (PE3) and (PE4) respectively. The bounds (RB2) and (RB3) require estimates of $\beta \cdot \sup_{y\in J-J'} \left|\frac{D_\mu g(u)}{D_x g(y)}\right|$. Note

$$\left|\frac{D_\mu g(y)}{D_x g(y)}\right| = \left|\frac{D_x h(f(y))D_\mu f(y)+D_\mu h(f(y))}{D_x h(f(y))D_x f(y)}\right| < (1 + 2E)\frac{|D_\mu f(u)|}{|D_x f(y)|}.$$

Moreover, $|D_\mu f(y)|$ is of order $\frac{\ell(\Gamma)}{|\mu_2 - \mu_1|}$ where Γ is the component of $I-\Lambda$ containing $f(0)$. Also, if u is the endpoint of Γ in f(J), $D_x h(f(u))$ is of order $\frac{\ell(J)}{\ell(\Gamma)}$ From (PE4), $|D_x h(f(u))||D_x f(u)| > \frac{1}{E} > 1$, so that $\beta\left|\frac{D_\mu g(y)}{D_x g(y)}\right|$ has order $\ell(J)$ or $S^{-1/2}$. Thus, when S is large (RB2) and (RB3) will be satisfied. For (RB4), (RB5) and (RB6), recall that $\alpha = \frac{\ell(J)}{2}$ is also of order $S^{-1/2}$. If D_1 is doubled, then the size of α and β need only be halved as well for (RB1)-(RB6) to continue to hold. The proposition follows since $S_0 > 2$.

There remain two parts of the proof of Jakobson's Theorem which must be completed. We need to (1) relate the parameter exclusions expressed in terms of values of f(0) to parameter values and (2) show that the recursion can be started with large values of S and fixed values for the constants D_i, E. We shall sketch these arguments, leaving some details to the reader.

Since the endpoints of I_j vary smoothly with μ, the Cantor sets $\Lambda_\mu = \{x|f_\mu^k(x) \in J \text{ for all } k \geq 0\}$ vary continuously with μ (indeed smoothly as we see below). As μ varies, the image $f_-(0)$ moves across the sets Λ_μ. The recursive hypotheses imply that $f_\mu(0)$ moves faster than points of Λ_μ and that the process happens with distortion that is controlled uniformly throughout the recursion. This leads to uniform bounds on the measures of the parameter intervals for wich the image of $f_\mu(0)$ lies in one of the excluded regions of I.

We can characterize the points of Λ symbolically: $f|_\Lambda$ is topologically equivalent to the full shift on the symbols $\{I_j\}$. If $\underline{a} = \{a_n\}$ is a sequence of indices and $\emptyset_n$ is the sequence of functions defined by $\emptyset_n = (f|_{I_{a_0}})^{-1^n} (f|_{I_{a_0}})^{-2} \cdots (f|_{I_{a_n}})^{-1}$, then $\emptyset_n(x,\mu)$ is a convergent sequence of functions with $|D_x\emptyset_n| < L^{-n}$ and $|D_\mu\emptyset_n| < \frac{1}{L-1}\sup|D_\mu f^{-1}|$. Thus the limit is a smooth function of rank 1 and points of Λ_μ vary smoothly with μ.

We assert that the lengths of the gaps Γ_μ that form the complement of Λ_μ vary smoothly with a uniform bound on the rate at which lengths can change. This bound comes from the second derivative estimates of (RH1). If Γ_μ is a gap of Λ_μ varying with μ, then there is a point $\xi_\mu \in \Gamma_\mu$ with $\ell(\Gamma_\mu)D_x h(\xi_\mu) = \ell(J_\mu)$. Now $\ell(J_\mu)$ changes at a bounded rate when it is not too small relative to $S^{-1/2}$ and $\log D_x$ has derivatives controlled by Lemma 2.3. These give a uniform bound on the variation of log $\ell(\Gamma_\mu)$, which is what was desired.

The parameter exclusions are formulated in terms of the position of $f_\mu(0)$ relative to Λ_μ. There are forbidden bands which consist of (1) gaps with short return times, and (2) bands near the endpoints of gaps Γ with width that is a small proportion of $\ell(\Gamma)$. All the excluded bands thus vary smoothly with μ, and there are uniform bounds on the logarithmic variation of their lengths. The nonlinear distortion of $f_\mu(0)$ as a function of μ is controlled by the constant D_6 which remains fixed throughout the recursion. Finally, the recursive hypothesis (RH3) implies that $|D_\mu f|$ is significantly larger than the rate at which endpoints of gaps move (if J_μ were fixed, this would be $\left|\frac{D_\mu h}{D_x h}\right|$. Therefore, the set of μ for which $f_\mu(0)$ is an excluded band has relative measure comparable to the measure of the exclude bands in I.

Finally, we can find parameter ranges in a smooth family for which it is possible to construct a renormalized induced mapping with specified values of the constants (D_i,E) and arbitrarily large S. If the family contains parameter values for which 0 has a trajectory which ends in an ustable periodic orbit γ (such as $f(x) = 2-x^2$), then nearby parameter values give induced maps for which the central branch spends arbitrarily long near the orbit γ. As the return time for 0 increases, S increases without bound. Moreover, the branches of the induced mapping spend most of their time before return near γ. The esimates of Lemma 2.3 imply that the distortion constants D_i of such a family remain bounded as the return time increases. Indeed, it is easily checked that in ranges of parameters with a restrictive point p_μ of period n, the renormalized mapping on the interval $[-p_\mu,p_\mu]$ differs from the family $(\mu-1)- \mu x^2$ by smaller and smaller distortion. This family satisfies all of the recursive hypotheses. Since the hypotheses describe an open set of strongly C^2 families (with respect to the obvious topology), one obtains a starting induced mapping for the recursion by first replacing f_μ with a renormalized induced mapping over a suitable parameter range and then forming its induced mappings.

REFERENCES

[1] M. Benedicks and L. Carleson, On the iterations of $1\text{-}ax^2$ on $(-1,1)$. Institut Mittag-Laffler, Report No. 3, 1983.

[2] P. Collet and J. P. Eckmann, On the abundance of aperiodic behaviour for maps of the interval, Comm. Math. Phys. 73 (1980), 115-160.

[3] J. Guckenheimer, Sensitive dependence to initial conditions for one dimensional maps, Comm. Math. Phys. 70 (1979), 133-160.

[4] M. Jakobson, Absolutely continuous invariant measures for one parameter families of one-dimensional maps. Comm. Math. Phys. 81 (1981), 39-88.

[5] Stewart Johnson, Thesis, Stanford University, 1985.

Contemporary Mathematics
Volume **58**, Part III, 1987

ORBITS AND HARMONIC MAPS

M.A. Guest[1]

ABSTRACT. A criterion is given for the projective weight orbit of a representation of a compact Lie group to be the image of a harmonic map.

1. INTRODUCTION. There is a well known relation between irreducible representations of a compact Lie group and certain algebraic homogeneous spaces of the group, developed by Borel-Weil [8], Borel-Hirzebruch [1], and Bott [2]. We shall examine a differential geometric aspect of this relation, which is relevant to work of Eells-Wood [3] on the construction of harmonic map between compact Kähler manifolds.

If G is a compact Lie grou, and $\theta : G \longrightarrow U_{n+1}$ is a homomorphism (i.e. a representation of G on $\mathbb{C}^{n+1}$, unitary with respect to the standard Hermitian form), there is an induced action of G on the projective space $\mathbf{P}(\mathbb{C}^{n+1}) = \mathbb{C}P^n$. If $v \in \mathbb{C}^{n+1}$ the orbit of the line $\langle v \rangle \in \mathbb{C}P^n$ will be denoted $G \cdot \langle v \rangle$. Let T be a maximal torus of G. The vector v is a *weight vector* for θ if the line $\langle v \rangle$ is fixed under the action of T, i.e. if $\langle v \rangle$ is itself a one-dimensional T-module. If λ is the associated character of T, the associated "infinitessimal character" Λ is called a *weight* of θ (Λ is a linear functional on the Lie algebra $L(T)$ of the torus and satisfies $\lambda_0 \exp = e^{2\pi i \Lambda}$). The weight vector v is said to have weight Λ, and the subspace of $\mathbb{C}^{n+1}$ consisting of vector which have weight Λ is called the *weight space* of Λ; the dimension of this weight space is the *multiplicity* of Λ. As θ is unitary, there exists a basis $v_0,\ldots,v_n$ of $\mathbb{C}^{n+1}$ consisting of weight vectors, whose weights will be written $\Lambda_0,\ldots,\Lambda_n$. The number of times a given weight is repeated in this list is equal to its multiplicity.

1980 Mathematics Subject Classification 53 C 30.

[1]Research supported in part by a grant from the National Science Foundation.
Lefschets Research Fellow 1982-3, Centro de Investigación y de Estudios Avanzados del I.P.N.

© 1987 American Mathematical Society
0271-4132/87 $1.00 + $.25 per page

If v is a weight vector of θ, we refer to $g \cdot \langle v \rangle \cong G/H$ as a *projective weight orbit*. If such an orbit is distinguished among orbits of G by the fact that the identity component of the stabilizer of $\langle v \rangle$ in the centralizer $C(S)$ of some subtorus S of T. Indeed, if

$$L(g) \otimes \mathbb{C} = L(T) \otimes \mathbb{C} \oplus \sum_{\alpha \in \Delta_G} E\alpha$$

is the complexified decomposition of the Lie algebra of G into weight spaces of the adjoint representation, the Lie algebra of H is given by

$$L(H) \otimes \mathbb{C} = L(T) \otimes \mathbb{C} \oplus \sum_{\alpha \in \Delta H} E\alpha$$

(These decompositions are known as root space decompositions; Δ_H denotes the roots of the group H. See [5] for further details).

The subset Δ_H of Δ_G is characterized by

$$\alpha \in \Delta_H \Longleftrightarrow E_{\pm\alpha} \cdot v = 0$$

where we use a dot to indicate the action of $L(g) \otimes \mathbb{C}$ on $\mathbb{C}^{n+1}$. (More precisely, to $\theta : G \to U_{n+1}$ there is associated a Lie algebra homomorphism $L(\theta): L(g) \otimes \mathbb{C} \to L(U_{n+1}) \otimes \mathbb{C}$. The linear operator on $\mathbb{C}^{n+1}$ corresponding to $L(\theta)\ x \in L(U_{n+1}) \otimes \mathbb{C} \cong \mathrm{End}(\mathbb{C}^{n+1})$ will be written $u \mapsto x \cdot u$, to simplify notation). These projective weight orbits are the subject of our study.

We summarize first some basic properties of "generalized flag manifolds", i.e. homogeneous spaces of the form $G/C(S)$, where $C(S)$ is the centralizer of a torus S. If $\Delta_{g/C(S)} (= \Delta_G - \Delta_{C(S)})$ denotes the set of complementary roots, a choice $\Delta^+_{G/C(S)} \subseteq \Delta_{G/C(S)}$ of positive roots defines a $C(S)$-invariant almost complex structure on $G/C(S)$, for which $(T_0 G/C(S))_{1,0}$ is identified with $\sum_{\alpha \in \Delta^+_{G/C(S)}} E_\alpha$. This is integrable if and only if $\alpha + \beta \in \Delta^+_{g/C(S)}$ for all $\alpha, \beta \in \Delta^+_{G/C(S)}$ such that $0 \neq \alpha + \beta \in \Delta_{G/C(S)}$. When $S = T$, $\Delta^+_{G/S(C)} = \Delta^+_G$, so a choice of positive roots of G defines a complex structure on G/T. All such choices give equivalent complex structures, and we fix one, denoted J, throughout this article. If $G = SU_n$, the roots with respect to the maximal torus $S(U_1 \times \ldots \times U_1)$ (diagonal matrices) may be written $\{x_i - x_j\}_{i \neq j}$, and we shall choose $\Delta^+_{SU_n} = \{x_i - x_j\}_{i > j}$. The negative of the Killing form of G gives a $C(S)$-invariant Riemannian metric $\langle , \rangle$ on $G/C(S)$, and any other $C(S)$-invariant metric differs from $\langle , \rangle$ only by a positive multiple on each irreducible $C(S)$-submodule of $T_0 G/C(S)$. For example, if $S = T$, $T_0 G/T \cong \sum_{\alpha \in \Delta^+} V_\alpha$ where $V_\alpha \otimes \mathbb{C} \cong E_\alpha \oplus E_{-\alpha}$, and each V_α is a two-dimensional (real) irreducible T-module. Thus, any T-invariant Riemannian metric on G/T is specified by a function $r : \Delta^+_G \to (0, \infty)$ and will be denoted $\langle , \rangle_r$, this being defined by $\langle x,y\rangle_r = r(\alpha)\langle x,y\rangle$ for all $x,y \in V_\alpha$. The metric $\langle\, ,\rangle_r$ is Kähler respect to the chosen complex- structure if an only if $r(\alpha) = \langle\alpha, \gamma\rangle$ for some linear functional $\gamma \in L(T)^*$ which is in the interior of the fundamental Weyl chamber defined by the choice of positive roots.

2. ORBITS. With the notation of the previous section, we consider $G\cdot\langle v\rangle$ as the image of the map $f\colon G/T \to \mathbb{C}P^n = SU_{n+1}/S(U_n \times U_1)$, *where* $S(U_n \times U_1)$ is the centralizer of $\langle v\rangle \subseteq \mathbb{C}^{n+1}$. This map is determined by its (complexified) derivative at $o \in g/T$

$$(df)_0\colon \sum_{\alpha\in\Delta_G} E_\alpha \to (T_0\mathbb{C}P^n)\otimes\mathbb{C} \cong \langle v\rangle^* \otimes \langle v\rangle^\perp \oplus (\langle v\rangle^\perp)^* \otimes \langle v\rangle$$

which is a homomorphism T-modules. A complex structure may be defined on $\mathbb{C}P^n$ by choosing $(\langle v\rangle^\perp)^* \otimes \langle v\rangle$ as $(T_0\mathbb{C}P^n)_{1,0}$. Thus, f is holomorphic if and only if $(df)_0 \sum_{\alpha\in\Delta_G} E_\alpha \subseteq (\langle v\rangle^\perp)^* \otimes \langle v\rangle$, i.e. if and only if $E_\alpha \cdot V = 0$ for all $\alpha \in \Delta_G^+$.

It is well known that if θ is irreducible then it has a unique maximal weight (of multiplicity one). If v is a weight vector for the maximal weight Λ, then $E_\alpha \cdot v = 0$ for all $\alpha \in \Delta_G^+$, since if $x \in E_\alpha$, $x\cdot v$ is either zero or a weight vector with weight $\Lambda + \alpha$. Hence the projective weight orbit is holomorphically embedded. In fact, $g\cdot\langle v\rangle$ is a (compact) orbit of the complexified group g^c.

We shall say that the projective weight orbit $g\cdot\langle v\rangle$ is W-*holomorphic* if for all $\alpha \in \Delta^+_{g\cdot\langle v\rangle}$, either $E_\alpha \cdot v \neq 0$ or $E_{-\alpha}\cdot v \neq 0$ (but not both). Thus, a W-holomorphic orbit acquires an almost complex structure (defined by $\{\alpha \in \Delta_{g\cdot\langle v\rangle} \mid E_\alpha \cdot v = 0\}$), with respect to which the embedding is holomorphic. For example, if Λ is in the Weyl group orbit of a weight whose projective weight orbit is holomorphic, then $g\cdot\langle v\rangle$ is W-holomorphic.

A W-holomorphic orbit is in fact symplectic, i.e. the restriction to $g\cdot\langle v\rangle$ of the symplectic form $w(x,y) = \langle x, Jy\rangle$ of $\mathbb{C}P^n$ is nondegenerate. Thus is a consequence of the following result of B. Kostant and S. Sternberg [7]. To state the result, we introduce some notation. It is well known that non-zero elements $x_\alpha \in E_\alpha$ may be chosen so that $\langle x_\alpha, x_{-\alpha}\rangle = 1$ and $\langle x_\alpha, x_\beta\rangle = 0$ if $\alpha + \beta \neq 0$ (see [5]). The element $h_\alpha = [x_\alpha, x_{-\beta}]$ is then in $L(T)\otimes\mathbb{C}$.

PROPOSITION. *The projective weight orbit* $G\cdot\langle v\rangle$ *is simplectic if and only if* $h_\alpha \cdot v \neq 0$ *for all* $\alpha \in \Delta_{g\cdot\langle v\rangle}$.

PROOF. It sufficies to check that the pull-back via $(df)_0$ of the form ω is nondegenerate. Since the pull-back is T-invariant, it sufficies to consider its restriction to each V_α. This is nondegenerate if and only if (its complexification) is non-zero on the pair $(x_\alpha, X_{-\alpha})$. We shall write x', x'' (respectively) for the elements of Hom $(\langle v\rangle^\perp, \langle v\rangle$, Hom $(\langle v\rangle, \langle v\rangle^\perp)$ given by $(df)_0 x$, for $x \in \sum_{\alpha \Delta_G} E\alpha : x'$ is the composition

$$\langle v\rangle^\perp \xrightarrow{x} \mathbb{C}^{n+1} \xrightarrow{pv} \langle v\rangle$$

and x'' is the composition

$$\langle v\rangle \xrightarrow{x} \mathbb{C}^{n+1} \xrightarrow{qv} \langle v\rangle^\perp$$

where pv, qv denote the orthogonal projections onto $\langle v\rangle, \langle v\rangle^\perp$. Since $\langle x, y\rangle$ is a non-zero multiple of $tr(x\,y)$ (i.e. the trace linear tranformation $w \longrightarrow x\cdot(\cdot w))$, the pull-back of ω to V_α is nondegenerate if and only if

$$tr(x'_\alpha + x''_\alpha)(x'_{-\alpha} - x''_{-\alpha}) \neq 0 .$$

As $x'_\alpha x'_{-\alpha}$ and $x''_\alpha x''_{-\alpha}$ are zero, this translates into the condition

$$tr(x'_\alpha x''_{-\alpha} - x'_{-\alpha} x''_\alpha) \neq 0$$

Now, $x'_\alpha x''_{-\alpha} - x'_{-\alpha} x''_\alpha$ is the scalar transformation $v \longrightarrow h_\alpha \cdot v$ of $\langle v\rangle$, so this completes the proof.

Note that

$$\begin{aligned}\|x_{-\alpha}\cdot v\|^2 - \|x_\alpha \cdot v\|^2 &= \langle x_{-\alpha}\cdot v, x_{-\alpha}\cdot v\rangle - \langle x_\alpha \cdot v, x_\alpha \cdot v\rangle \\ &= \langle x_\alpha x_{-\alpha}\cdot v, v\rangle - \langle x_{-\alpha} x_\alpha \cdot v, v\rangle \\ &= \langle h_\alpha \cdot v, v\rangle ,\end{aligned}$$

hence $h_\alpha \cdot v \neq 0$ if and only if $\|x_\alpha \cdot v\| \neq \|x_{-\alpha}\cdot v\|$. The fact that a W-holomorphic orbit is symplectic follows from this.

It follows from a well known result of Kostant and Sourian that a simplectic orbit $G\cdot\langle v\rangle$ covers the orbit of the weight Λ of v under the co-adjoint action of G, i.e. $G/C(S)$ where S is the torus generated by Λ. If $\pi_1 G = 0$, it follows that $\pi_1 G/C(S) = 0$, hence $G\cdot\langle v\rangle \cong G/C(S)$.

Example: $G = SU_{k+1}$. The weights of SU_{k+1} with respect to the standard maximal torus $T = S(U_1 x \ldots x U_1)$ are the linear functionals $\sum_{i=0}^{k} n_i x_i \in L(T)^*$, where $n_i \in \mathbb{Z}$ and $\sum_{i=0}^{k} x_i = 0$. We choose $\{x_i - x_j\}_{0 \leqslant j < i \leqslant n}$ to be the set of positive roots, as usual. Up to a non-zero scalar multiple, we may take

$x_\alpha \in E_\alpha$ to be the $(k+1) \times (k+1)$ matrix $E_{i,j}$ which has a 1 in the $(i+1, j+1)$-th position and zeros elsewhere, where $\alpha = x_i - x_j$. Let $\sigma : SU_{k+1} \longrightarrow SU_{k+1}$ be the identity representations of SU_{k+1} on $\mathbb{C}^{k+1}$. We Shall examine various representations θ which are constructed from σ.

<u>1. $\theta = S^n\sigma (n \geqslant 1)$</u>. If $e_0, e_1, \ldots, e_k$ is an orthonormal basis for $\mathbb{C}^{k+1}$, we choose an isometric isomorphism $S^n(\mathbb{C}^{k+1}) \cong \mathbb{C}^N$, $N = \binom{n+k}{n}$, by requiring the monomials $\prod_{i=0}^{k} e_i^{\ell_i} (n!/\ell_0! \ldots \ell_k!)^{\frac{1}{2}}$ to be an orthonormal basis of $S^n(\mathbb{C}^{k+1})$ $(\sum_{i=0}^{k} \ell_i = n, \ell_i \geqslant 0)$. The natural representation of SU_{k+1} on $\mathbb{C}^N$ is then unitary, and is well known to be irreducible. The monomials are in fact weight vectors, with weights $\sum_{i=0}^{k} \ell_i x_i$, all of which have multiplicity one. To calculate the action of the elements $E_{i,j}$ on weight vectors, note that

$$x \cdot v = (d/dt)\, \theta\, (\exp tx)\, v \Big|_{t=0}$$

for any $x \in L(U_{k+1}) \otimes \mathbb{C} \cong \mathrm{End}\,(\mathbb{C}^{k+1})$, where exp denotes exponentiation of matrices and θ is extended to the complexified group $U_{k+1}^{\mathbb{C}} = G\ell_{k+1}\mathbb{C}$. Thus

$$E_{i,j} \cdot e_0^{\ell_0} \ldots e_k^{\ell_k} = (d/dt)\;\; \theta\, (I + tE_{i,j})\, e_0^{\ell_0} \ldots e_k^{\ell_k} \Big|_{t=0}$$

$$= (d/dt)\, e_0^{\ell_0} \ldots (e_j + te_i)^{\ell_j} \ldots e_k^{\ell_k} \Big|_{t=0}$$

$$= \ell_j\, e_0^{\ell_0} \ldots e_j^{\ell_j - 1} \ldots e_i^{\ell_i + 1} \ldots e_k^{\ell_k} \quad .$$

Hence $E_{i,j} \cdot e_0^{\ell_0} \ldots e_k^{\ell_k} = 0$ if and only if $\ell_j = 0$. The maximal weight is nx_k, and the corresponding projective weight orbit has $\{\pm(x_k - x_i)\}_{0 \leqslant i \leqslant k-1}$ as its set of complementary roots. It is isomorphic to $\mathbb{C}P^k$, and the embedding

$$\mathbb{C}P^k \longrightarrow \mathbb{C}P^N, \quad N = \binom{n+k}{n}$$

may be written explicitly homogeneous coordinates as

$$[z_0; \ldots; z_k] \longrightarrow [\ldots, z_0^{\ell_0} \ldots z_k^{\ell_k} (n!/\ell_0! \ldots \ell_k!), \ldots]$$

This is (projectively equivalent to) the classical Veronese embedding. The projective weight orbit for $\sum_{i=0}^{k} \ell_i x_i$ is locally isomorphic to $SU_{k+1}/S(U_r \times U_1 \times \ldots \times U_1)$ ($k+1-r$ factors equal to U_1), where r is the number of integers ℓ_i which are equal to zero. Such an orbit is W-holomorphic if and only if $r = k$, i.e. for the weights $nx_0, \ldots, nx_k$ (which constitute the Weyl group orbit of the maximal weight).

Since

$$[E_{i,j}, E_{j,i}] \cdot e_0^{\ell_0} \ldots e_k^{\ell_k} = (\ell_i(\ell_{j+1}) - \ell_j(\ell_{i+1}))e_0^{\ell_0} \ldots e_k^{\ell_k}$$
$$= (\ell_i - \ell_j)e_0^{\ell_0} \ldots e_k^{\ell_k} ,$$

the orbit is symplectic if and only if the non-zero integers ℓ_i are all distinct. Thus, there are many non-symplectic orbits in general

2. <u>$\theta = \Lambda^n \sigma$ $(1 \leqslant n \leqslant k+1)$</u> . A similar analysis applies to this irreducible representation. The indecomposable elements $e_0^{\ell_0} \wedge \ldots \wedge e_k^{\ell_k}$ are weight vectors with weights $\sum_{i=0}^{n} \ell_i x_i$ ($\ell_i = 0$ or 1, $\sum_{i=0}^{k} \ell_i = n$), all of which have multiplicity one. The maximal weight is $x_{k-n+1} + \ldots + x_k$, and the corresponding projective weight orbit gives the standard Plücker embedding of the Grassmannian $Gr_n(\mathbb{C}^{k+1})$. All other weights are in the same Weyl group orbit.

3. <u>$\theta = \mathrm{Ad} \otimes \mathbb{C} \cong \sigma \otimes \bar{\sigma} - 1$</u> . As a basis for the representation space $L(U_{k+1}) \otimes \mathbb{C} \cong \mathrm{End}\ (\mathbb{C}^{n+1})$ we may take the elements $E_{i,j} (0 \leqslant i,j \leqslant k)$. The weights are the roots, together with zero, the latter being the only one not to have multiplicity one (it has multiplicity k) . The maximal weight is $x_k - x_0$, and the corresponding projective weight orbit is isomorphic to $SU_{k+1}/S(U_1 \times U_1 \times U_{k-1})$.

All other non-zero weights are in the same Weyl group orbit. For the zero weight, the projective weight orbits are locally isomorphic to the generalized flag manifolds $SU_{k+1}/S(U_{n_1} \times \ldots \times U_{n_r})$ where $n_1 + \ldots n_r = n$, none of which are symplectic.

3. <u>HARMONIC MAPS</u>. A map $f : M \longrightarrow N$ of Riemannian manifolds is said to be *harmonic* if $\mathrm{tr}\ \nabla(df) = 0$ (the tensor $\nabla(df) \in \Gamma\ (S^2T^*M \otimes f^{-1}TN)$ is the *second fundamental form* of f, and $\mathrm{tr}\ \nabla(df)$ is its contraction with respect to the metric of M). If $M = G/H$ is a Riemannian homogeneous space, the covariant derivative operator in TM is determined by an H-module transformation $\Lambda : \otimes^2 T_o M \longrightarrow T_o M$ (see [6], volume II, chapter 10). The following is proved

in [4] .

LEMMA. *Let G/H, G'/H' be Riemannian homogeneous spaces with covariant derivative operators given by transformations Λ, Λ'. Let $f : G/H \longrightarrow G'/H'$ be induced by a homomorphism $\theta : G \longrightarrow G'$ (thus $\theta(H) \subseteq H'$). Then the second fundamental form $\nabla(df)$ is determined by the H - module transformation*

$$B : S^2(T_o G/H) \longrightarrow T_o G'/H'$$

which is given in terms of the induced homomorphism $L(\theta) : L(G) \longrightarrow L(G')$ of Lie algebras by the formula

$$B(x,y) = [(L(\Theta)x)_{H'} \quad , L(\Theta)y)_{m'} \quad] \; +$$

$$\Lambda'((L(\Theta)x)_{m'} \; , (L(\Theta)y)_{m'}) - \theta(\Lambda(x,y))_{m'}$$

$$(x,y \in m = T_o G/H \; , m' = T_o G'/H' , \; H' = L(H')) \; .$$

We shall apply this result to the map $f : G/T \longrightarrow \mathbb{C}P^n$ introduced in section 2. As T-modules,

$$(T_o G/T) \otimes \mathbb{C} \cong \sum_{\alpha \in \Delta} E_\alpha \quad \text{and}$$

$$(T_o \mathbb{C}P^n) \otimes \mathbb{C} \cong \sum_{0 \leqslant i \neq j \leqslant n} \rho^{i,j} \quad ,$$

where $\rho^{i,j}$ is the one-dimensional module whose infinitessimal character is $\Lambda_i - \Lambda_j$. Even without the formula of lemma, the fact that B is a T-module transformation has certain implications. For example, if the representation θ has no repeated weights (i.e. $\Lambda_i \neq \Lambda_j$ if $i \neq j$), then $(T_o\mathbb{C}P^n) \otimes \mathbb{C}$ contains no trivial submodule (of positive dimension). Since $\operatorname{tr} \nabla(df)$ is essentially the value of B on the trivial submodule of $S^2(T_o G/T)$ given by the metric, it must therefore be zero. We have seen examples of such representations in section 2, so there give a large class of harmonic maps. More generally, one has the following criterion.

THEOREM. *Let G/T have any T - invariant metric $\langle , \rangle_r$, and let $\mathbb{C}P^n$ have its standard (Fubini-Study) metric. Let $\theta : G \longrightarrow SU_{n+1}$ be a representation of G. If $v \in \mathbb{C}^{n+1}$ is a weight vector of θ, the map*

$$f : G/T \longrightarrow \mathbb{C}P^n , \quad gT \longrightarrow g \cdot \langle v \rangle$$

is harmonic if and only if v *is an eigenvector of the operator*

$$\sum_{\alpha \in \Delta_G^+} (1/r(\alpha)) \; X_\alpha \; X_{-\alpha}$$

PROOF. As an orthonormal basis (over $\mathbb{R}$) for $T\,G/T \subseteq (T\,G/T) \otimes \mathbb{C}$, one may take the vectors $(1/\sqrt{2r(\alpha)})\, i(x_\alpha - x_{-\alpha})$, $(1/\sqrt{2r(\alpha)})(x_\alpha + x_{-\alpha})$, $\alpha \in \Delta_G^+$. From this it follows that f is harmonic if and only if the element $\sum_{\alpha \in \Delta_G^+} (1/r(\alpha))(x_\alpha \otimes x_{-\alpha} + x_{-\alpha} \otimes x_\alpha)$ is in the kernel of the map B of the lemma. It may be shown directly from the definition of Λ that $\Lambda(x_\alpha, x_{-\alpha}) = 0$ for all $\alpha \in \Delta$, and that $\Lambda' = 0$ (see [4]).

Hence the required condition is that

$$\sum_{\alpha \in \Delta_G^+} (1/r(\alpha)) \, (L(\theta)x_\alpha \otimes L(\theta)x_{-\alpha} + L(\theta)x_{-\alpha} \otimes L(\theta)x_\alpha)$$

be in the Kernel of the map

$$\otimes^2 (L(SU_{n+1}) \otimes \mathbb{C}) \longrightarrow L(SU_{n+1}) \otimes \mathbb{C}$$

$$(x,y) \longrightarrow [x_{\mathfrak{h}'} , y_{\mathfrak{m}'}]$$

where $\mathfrak{h}' = L(S(U_n \times U_1)) \otimes \mathbb{C}$, $\mathfrak{m}' = (T_0\,\mathbb{C}P^n) \otimes \mathbb{C}$. In other words, $z = \sum_{\alpha \in \Delta_G^+} L(\theta)x_\alpha \otimes L(\theta)x_{-\alpha}$ is required to be in the kernel of the map

$$(*) \qquad (x,y) \longrightarrow [x_{\mathfrak{h}'} , y_{\mathfrak{m}'}] - [x_{\mathfrak{m}'} , y_{\mathfrak{h}'}] .$$

We may write

$$L(SU_{n+1}) \otimes \mathbb{C} = L(S(U_1 \times \ldots \times U_1)) \otimes \mathbb{C} \oplus \sum_{0 \leqslant i \neq j \leqslant n} \rho^{i,j} ,$$

and assume without loss of generality that $v = v_0$, $\langle v_1, \ldots, v_n \rangle = \langle v \rangle^\perp$. Then $L(\theta)x_\alpha \in \sum_{0 \leqslant i \neq j \leqslant n} \rho^{i,j}$, and the component $(L(\theta)x_\alpha)_{i,j}$ in $\rho^{i,j} \cong \mathrm{Hom}(\langle v_j \rangle, \langle v_i \rangle)$ is just the map

$$\langle v_j \rangle \xrightarrow{x_\alpha} \mathbb{C}^{n+1} \xrightarrow{p_i} \langle v_i \rangle$$

where p_i denotes orthogonal projection onto $\langle v_i \rangle$. It is clear that

$L(\theta)x_\alpha$ lies entirely in $\sum_{0\leqslant i\neq j\leqslant n} \rho^{i,j}$, so for our purposes we may restrict the map (*) to this subspace. As a transformation of T-modules, the restriction may be identified with the composition

$$\otimes^2 \sum_{0\leqslant i\neq j\leqslant n} \rho^{i,j} \longrightarrow \sum_{0\leqslant i\neq j\leqslant n} \rho^{i,j} \xrightarrow{(\pi_{1,0},-\pi_{0,1})} \sum_{i=1}^{n} \rho^{i,0} \oplus \sum_{i=1}^{n} \rho^{0,i}$$

where the first map is the natural one which identifies $\rho^{i,j} \otimes \rho^{j,k}$ and $\rho^{j,k} \otimes \rho^{i,j}$ with $\rho^{i,k}$, and the second is given by the appropiate projections. The image of $L(\theta)x_\alpha \otimes L(\theta)x_{-\alpha}$ in $\sum_{i=1}^{n} \rho^{i,0} \cong \langle v\rangle^* \otimes \langle v\rangle^\perp \cong \mathrm{Hom}(\langle v\rangle, \langle v\rangle^\perp)$ is the map

$$\langle v\rangle \xrightarrow{x_\alpha x_{-\alpha} + x_{-\alpha} x_\alpha} \mathbb{C}^{n+1} \xrightarrow{q_v} \langle v\rangle^\perp .$$

The image in $\sum_{i=1}^{n} \rho^{0,1} \cong (\langle v\rangle^\perp, \langle v\rangle)$ is the map

$$\langle v\rangle^\perp \xrightarrow{-(x_\alpha x_{-\alpha} + x_{-\alpha} x_\alpha)} \mathbb{C}^{n+1} \xrightarrow{pv} \langle v\rangle .$$

Hence the condition for harmonicity is that the operator $\sum_{\alpha\in\Delta_g^+} (1/r(\alpha))(x_\alpha x_{-\alpha} + x_{-\alpha} x_\alpha)$ should preserve the decomposition $\mathbb{C}^{n+1} = \langle v\rangle \oplus \langle v\rangle^\perp$. Since the operator $\sum_{\alpha\in\Delta_g^+} (1/r(\alpha))(x_\alpha x_{-\alpha} - x_{-\alpha} x_\alpha) = \sum_{\alpha\in\Delta_g^+} (1/r(\alpha))h_\alpha$ preserves the decomposition, and since all operators here are Hermitian, the result follows.

Observing that $x_\alpha x_{-\alpha} \cdot v$ is either zero or a weight vector with the same weight Λ as v, we see that each operator $x_\alpha x_{-\alpha}$ preserves each one-dimensional weight space. Thus, the map f is certainly harmonic if Λ has multiplicity one (c.f. the remarks at the beginning of this section). A less obvious corollary is that if θ is irreducible, f is harmonic for *every* weight vector v providing the Killing metric $\langle\,,\rangle$ is used on G, since $\sum_{\alpha\in\Delta_g^+} x_\alpha x_{-\alpha}$ has the same eigenvalues as the Casimir operator of θ, which in this case is a scalar transformation.

REMARKS.

1. Given the representation $\theta : G \longrightarrow U_{n+1}$, there is an induced action of G on any Grassmannian $Gr_m(\mathbb{C}^{n+1})$ (m-planes in $\mathbb{C}^{n+1}$), or more generaly on

any flag manifold $F_{m_1,\dots,m_k}(\mathbb{C}^{n+1})$ (sequences $E_{m_1} \subseteq E_{m_2} \subseteq \dots \subseteq E_{m_k} = \mathbb{C}^{n+1}$ where E_i is an i-plane in $\mathbb{C}^{n+1}$). The theorem may be extended to give a criterion for harmonicity of an orbit in this situation. An important example where such orbits arise naturally is given by the osculating bundles associated to a projective weight orbit. For these results, see [4] .

2.- The osculating bundle construction may be illustrated by the example $G = SU_2$, $\theta = S^n\sigma$. If $f : \mathbb{C}P^1 \longrightarrow \mathbb{C}P^n$ is the inclusion of the maximal projective weight orbit (i.e. a "rational normal curve of degree n" , c.f. example 1 in section 2), one has *associated curves*.

$$f_i : \mathbb{C}P^1 \longrightarrow Gr_{i+1}(\mathbb{C}^{n+1}) \quad (i = 0,1,\dots)$$

defined locally using homogeneous coordinates by

$$f_i(z) = f(z) \wedge f^1(z) \wedge \dots \wedge f^{(i)}(z) .$$

The curves are globally defined holomorphic curves; indeed, the composition of f_i with the Plücker embedding $Gr_{i+1}(\mathbb{C}^{n+1}) \longrightarrow \mathbb{C}P^N$ $(N = \binom{n+1}{i+1} - 1)$ is a projective weight orbit of the representation $\Lambda^{i+1}S^n\sigma$. (The relevant weight is the maximal weight of the largest summand $S^{(i+1)(n-i)}\sigma$ of $\Lambda^{i+1}S^n\sigma$) . If the weights of $\theta = S^n\sigma$ are arranged in descending order, let $V_0,\dots,V_n$ be corresponding weight vectors, so that $SU_2 \cdot \langle v_0 \rangle = f(\mathbb{C}P^1)$. Then $SU_2 \cdot \langle v_i \rangle = f_{(i)}(\mathbb{C}P^1)$, where

$$f_{(i)} : \mathbb{C}P^1 \longrightarrow \mathbb{C}P^n \quad (i = 0,\dots,n)$$

is defined by

$$f_{(i)}(z) = f_{i-1}(z)^{\perp} \cap f_i(z)$$

Thus, the holomorphic map f gives rise to harmonic "transforms" $f_{(0)}$ $(=f)$, $f_{(1)},\dots,f_{(n)}$. The W-holomorphic projective weight orbits are $SU_2 \cdot \langle v_0 \rangle$ and $SU_2 \cdot \langle v_n \rangle$; although these represent the same subset of $\mathbb{C}P^n$, they are embedded holomorphically and antiholomorphically by $f_{(0)}$ and $f_{(n)}$ (respectively) . The harmonic map $f_{(i)}$ may be considered as a combination of the extreme cases $f_{(0)}, f_{(n)}$, and indeed each of these may be recovered from $f_{(i)}$ by applying the osculating bundle construction. This point of view was taken in [3] where it was shown that $h_{(i)}$ is harmonic for *any* holomorphic map $h : \mathbb{C}P^1 \to \mathbb{C}P^n$, and that *all* harmonic maps $\mathbb{C}P^1 \to \mathbb{C}P^n$ are of the form $h_{(i)}$ (to suitable h and i). With hindsight, it may now be seen that the projective weight orbits of the irreducible representation of SU_2 generate all harmonic maps from $\mathbb{C}P^1$ to complex projective space, al follows. First, it is elementary that any holomorphic map $h : \mathbb{C}P^1 \longrightarrow \mathbb{C}P^n$ is of the form P.f , where $f : \mathbb{C}P^1 \longrightarrow \mathbb{C}P^n$ is the embedding of the maximal projective weight orbit of $S^n\sigma$ described above, and P is a (possibly improper) projectivity of $\mathbb{C}P^n$. By linearity, $(P \cdot f)_i = P \cdot f_i$, hence

$$h_{(i)}(z) = P \cdot f_{i-1}(z)^{\perp} \cap P \cdot f_i(z)$$

so $h_{(i)}$ is obtained explicity from the projective weight orbits of $S^n\sigma$.

BIBLIOGRAPHY

1. Borel, A., Hirzebruch, F., "Characteristic classes and homogeneous spaces I", Amer. J. Math. 80 (1985), 458-538.

2. Bott, R., "Homogeneous vector bundles", Annals of Math. 66 (1957), 203-248.

3. Eells, J., Wood, J.C., "Harmonic maps from surfaces to complex projective spaces", Advances in Math. 49 (1983), 217-263.

4. Guest, M.A., "Geometry of maps between generalized flag manifolds", to appear.

5. Helgason, S., Differential Geometry, Lie Groups, and Symmetric Spaces, Academic Press, New York, 1978.

6. Kobayashi, S., Nomizu, K., Foundations of Differential Geometry, Wiley, New York, volume I 1963, volume II 1969.

7. Kostant, B., Sternberg, S., "Symplectic projective orbits", in: New Directions in Applied Mathematics (P. Hilton and G.S. Young, editors), Springer-Verlag, New York, 1982, 81-84.

8. Serre, J.-P., "Reprèsentations Lineàires et espaces homogènes Kählériens des groupes de Lie compacts", Séminaire Bourbaki 100 (1954)

Department of Mathematics
Arkansas State University
AR 72467, U.S.A.

Current Address:
Department of Mathematics
Tulane University
New Orleans, LA 70118
U.S.A.

Contemporary Mathematics
Volume 58, Part III, 1987

SOME EXAMPLES OF INFINITE DIMENSIONAL SYSTEMS

Jack K. Hale

ABSTRACT. A class of infinite dimensional systems is described for which there exists a compact attractor. By restricting the flow to the attractor one can begin to develop a qualitative theory. Examples are given illustrating how this approach is leading to new results in partial differential equations and delay equations.

1. INTRODUCTION. Many applications involve dynamical systems in infinite dimensional spaces; for example, those generated by partial differential equations and delay differential or functional differential equations. Due to the complexities involved in performing analysis in infinite dimensions, one often imposes simplifying assumptions on the model to obtain a problem in n dimensional euclidean space R^n or, more generally, in a finite dimensional manifold M^n. These assumptions can be either physical hypotheses or some finite dimensional approximation scheme. Using the extensively developed geometric theory of dynamical systems on a finite dimensional manifold, one attempts to understand the behavior of the original system as the relevant physical parameters are subjected to variations.

The validity of the process of reduction to some finite dimensional space can only be justified if we have some reasonable geometric theory for infinite dimensional dynamical systems themselves. That is, we must have a theory for infinite dimensional systems which uses and appropriately modifies the concepts and ideas from finite dimensions in a significant way and, at least on a theoretical level, gives a prescription for determining the nature of the dynamics. Also, a dynamical system being infinite dimensional does not necessarily mean that it can exhibit every type of dynamics that occurs in ordinary differential equations in R^n. The infinite dimensional system may have special properties which must be isolated a priori and will determine the types of approximating systems that are feasible.

Although efforts are being made to extend the finite dimensional ideas

---

1980 Mathematics Subject Classification: 34C35, 34D30, 47H20, 35B40.

[1] Supported by the Air Force Office of Scientific Research under grant No. 84-0376; by the U.S. Army Research Office under contract No. DAAG-29-83-K-0029; and by the National Science Foundation under grant No. DMS-8205355.

© 1987 American Mathematical Society
0271-4132/87 $1.00 + $.25 per page

to infinite dimensions, the global theory is still in its infancy. One reason the development has been so slow certainly is due to the complexities that are introduced by the infintie dimensionality. It is difficult to isolate important examples which can be analyzed in sufficient detail to bring out some of the new phenomena that occur and, at the same time, show the advantages to be gained from considering the system as an abstract dynamical system. These advantages can be either in the discovery of new types of behavior or, perhaps more importantly, the approach can lead to the formulation and anlaysis of much more meaningful problems.

One purpose of these notes is to give some illustrations of infinite dimensional problems for which the techniques and ideas of dynamical systems have led to surprisingly new results in classical problems, to improvements of results that have been obtained by more classical approaches and to the formulation of more meaningful and significant problems.

The outline of the paper is as follows. In Section 2, we isolate a class of semigroups, the so called asymptotically smooth semigroups, for which there is maximal compact invariant set (a compact attractor) A which contains the ω-limit set of any bounded set. This shows that the essential part of the flow is the flow obtained by restricting the semigroup to the attractor. In Section 3, we discuss in some detail the simplest semigroups; namely, the gradient systems. One of the important results here is to give conditions under which the attractor is upper-semicontinuous with respect to parameters defining the semigroup. This gives results on structural stability.

Section 4 is devoted to examples of various types of evolutionary equations that lead to asymptotically smooth semigroups, together with sepcific examples where the approach of dynamical systems has "paid off". These examples deal with parabolic systems, retarded functional differential equations, and linearly damped nonlinear wave equations.

2. ASYMPTOTICALLY SMOOTH DYNAMICAL SYSTEMS AND COMPACT ATTRACTORS. Let X be a Banach space, $R^+ = [0,\infty)$. Suppose $T(t): X \to X$, $t \geq 0$, is a C^r-semigroup, $r \geq 0$; that is, $T(t)$, $t \geq 0$ is a semigroup with $T(t)x$ continuous in t, x together with derivatives in x up through order r. Let $\gamma^+(x) = \{T(t)x, t \geq 0\}$ be the <u>positive orbit</u> through x. A <u>negative orbit</u> through x is a function $\phi: (-\infty,0] \to X$ such that $\phi(0) = x$ and for any $s < 0$, $T(t)\phi(s) = \phi(t+s)$, $0 \leq t \leq -s$. A <u>complete orbit</u> through x is a function $\phi: R \to X$ such that $\phi(0) = x$ and $T(t)\phi(s) = \phi(t+s)$ for all $t \geq 0$, $s \in R$. To say there is a negative or complete orbit through x may impose restrictions on x since $T(t)$ is only a semigroup. Also, since $T(t)$ may not be $1 - 1$, there may be more than one negative orbit through x. We let $\gamma^-(x)$ [$\gamma(x)$] be the union of all negative orbits [complete] through x. If $B \subset X$ is a set we define $\gamma^+(B), \gamma^-(B), \gamma(B)$

as the union of the corresponding orbits for each point in B. The <u>ω-limit set</u> of B, $\omega(B)$, is defined as

$$\omega(B) = \cap_{s \geq 0} C\ell\ \gamma^{+}(T(s)B)$$

Similarly, one defines the <u>α-limit set</u> $\alpha(B)$. A set $S \subset X$ is said to be <u>invariant</u>, if, for any $x \in S$, there is a complete orbit $\gamma(x)$ through x and $\gamma(x) \in S$. This is equivalent to $T(t)S = S$ for $t \geq 0$.

If $C\ell\ \gamma^{+}(B)$ is nonempty and compact, then $\omega(B)$ is nonempty, compact and invariant. In addition, B connected implies $\omega(B)$ connected. Even if $C\ell\ \gamma^{+}(B)$ is not compact, the set $\omega(B)$ is invariant. If $\omega(B)$ is compact, then $\mathrm{dist}(T(t)B, \omega(B)) \to 0$ as $t \to \infty$. Similar remarks hold for the α-limit sets. The proofs of these results are standard.

A set $B \subset X$ is said to <u>attract a set</u> $C \subset X$ <u>under</u> $T(t)$ if $\mathrm{dist}(T(t)C, B) \to 0$ as $t \to \infty$. The semigroup $T(t)$ is <u>asymptotically smooth</u> if, for any bounded set $B \subset X$, there is a compact set $J = J(B) \subset X$ such that, J attracts the set $L(B) = \{x \in B\colon T(t)x \in B$ for $t \geq 0\}$. Special cases of asymptotically smooth maps are (i) there is a $t_1 \geq 0$ such that $T(t)$ is completely continuous for $t > t_1$, (ii) $T(t) = S(t) + U(t)$ where $U(t)$ is completely continuous for $t > 0$ and $S(t)$ is a linear mapping for which there is a constant $\alpha > 0$ such that $|S(t)| \leq \exp(-\alpha t)$, $t \geq 0$. For a proof, see [9]. These special cases are very important in the applications, but more general semigroups must be considered.

Asymptotically smooth maps were introduced by Hale, LaSalle and Slemrod [8] (see also Hale and Lopes [9]). It was shown that an asymptotically smooth map for which there is a compact set that attracts arbitrary compact sets of X must have a maximal compact invariant set that attracts a neighborhood of any compact set. Cooperman [5] and Massatt [20] clarified the attractivity properties of maximal compact invariant sets. These latter results led to a more applicable test for the existence of a compact attractor. The following result is very useful in the applications and the proof may be supplied as in the one of Massatt [20] for α-contractions. A semigroup $T(t)$ is <u>point dissipative</u> if there is a bounded set $U \subset X$ such that U attracts points of X.

<u>THEOREM 2.1</u>. <u>If</u> $T(t)$, $t \geq 0$, <u>is asymptotically smooth, point dissipative and orbits of bounded sets are bounded, then there exists a compact attractor for</u> $T(t)$.

If the semigroup $T(t)$ is a linear contraction plus a completely continuous map, then it is known from the work of Mallet-Paret [17] and Mañé [19] that the compact attractor A has finite capacity; in particular, finite Hausdorf dimension. It is not known if this is true for asymptotically smooth maps.

3. GRADIENT SYSTEMS. In the previous section, we have given a general class of semigroups for which there is a compact attractor. The basic problem remaining is to discuss the flow on the attractor and how the flow changes as one changes physical parameters in the problem that defines the semigroup. This problem of course is very difficult and one must restrict the discussion to special cases. As in ode's one begins with the simplest cases; namely, gradient systems.

Following Hale [6], we say a strongly continuos C^r-semigroup $T(t): X \to X$, $t \geq 0$, $r \geq 2$, is said to be a gradient system if

(i) Each bounded positive orbit is precompact

(ii) There exists a Liapunov function for $T(t)$; that is, there is a continuous function $V: X \to \mathbb{R}$ with the property that

(ii_1) $V(x)$ is bounded below,

(ii_2) $V(x) \to \infty$ as $|x| \to \infty$,

(ii_3) $V(T(t)x)$ is nonincreasing in t for each x in X,

(ii_4) if $V(T(t)x) = V(x)$ for $t \geq 0$, then x is an equilibrium point.

Let E be the set of equilibrium points of $T(t)$.

LEMMA 3.1. *If $T(t)$ is a gradient system, then the ω-limit set $\omega(x)$ for each $x \in X$ belongs to E. If $\gamma^-(x)$ is a precompact orbit through x, then the α-limit set $\alpha(x)$ of x belongs to E.*

Following the proof of a slightly less general result in [6], one obtains

THEOREM 3.2. *Suppose $T(t)$ is an asymptotically smooth gradient system for which orbits of bounded sets are bounded. If E is a bounded set, then there is a compact attractor A. If, in addition, each element of E is hyperbolic, then E is a finite set, $\dim W^u(\phi) < \infty$ and*

$$A = \cup_{\phi \in E} W^u(\phi)$$

where $W^u(\phi)$ is the unstable set of ϕ.

Let us now consider a family of gradient semigroups $T_\lambda(t)$, $t \geq 0$, on X, where λ is a parameter in a subset of a Banach space and $T_\lambda(t)x$ is continuous in (t,x,λ) and C^r in x for some $r \geq 1$. Let E_λ be the set of equilibrium points of $T_\lambda(t)$ and let A_λ be the attractor of $T_\lambda(t)$.

For any two subsets A,B of X, let

$$\delta(B,A) = \sup_{x \in B} \inf_{y \in A} |x-y|$$

A family of subsets B_λ, $\lambda \in \Lambda$, of X is said to be upper semicontinuous at λ_0 [lower semicontinuous at λ_0] if $\delta(B_\lambda, B_{\lambda_0})[\delta(B_{\lambda_0}, B_\lambda)] \to 0$ as $\lambda \to \lambda_0$. It is said to be continuous at λ_0 if it is both upper and lower semicontinuous at λ_0.

THEOREM 3.3. *Suppose* Λ *is a subset of a Banach space and, for each* $\lambda \in \Lambda$, $T_\lambda(t)$, $t \geq 0$, *is a* C^r*-gradient semigroup on* X, $r \geq 1$, *and* $T_\lambda(t)x$ *is continuous in* (t,x,λ) *and continuous in* λ *uniformly with respect to* (t,x) *in bounded sets. Suppose also that* $T_\lambda(t)$, *for each* $\lambda \in \Lambda$, *is asymptotically smooth and orbits of bounded sets are bounded. If* E_λ *is a bounded set for each* $\lambda \in \Lambda$ *and upper semicontinuous at* λ_0, *then the attractor* $\mathscr{A}_\lambda$ *is compact, attracts bounded sets of* X *under* $T_\lambda(t)$ *and* $\mathscr{A}_\lambda$ *is upper semicontinuous at* λ_0.

Theorem 3.3 says that the "size" of the attractor $\mathscr{A}_\lambda$ cannot increase much for λ near λ_0 if the hypotheses of the theorem are satisfied. It may, however, become smaller unless additional conditions are imposed on the flow on $\mathscr{A}_{\lambda_0}$. The appropriate conditions to impose on this flow are concerned with the stable and unstable sets of the equlibrium points of $T_{\lambda_0}(t)$. To state a general result, we need more notation. If $\phi \in E_\lambda$, let

$$W^s_\lambda(\phi) = \{x \in X: T_\lambda(t) \to \phi \text{ as } t \to \infty\}$$

the stable set of ϕ, and let $W^u_\lambda(\phi)$ be the unstable set of ϕ. If ϕ is hyperbolic, then $W^s_{\lambda,loc}(\phi) = \{x \in W^s_\lambda(\phi): \gamma^+(x) \subset U\}$, U a neighborhood of ϕ, is a submanifold of codimension equal to $\dim W^u_\lambda(\phi)$ at ϕ.

We say $T_\lambda(t)$ is *equivalent* to $T_{\lambda_0}(t)$, $T_\lambda(t) \sim T_{\lambda_0}(t)$, if there is a homeomorphism $h: \mathscr{A}_\lambda \to \mathscr{A}_{\lambda_0}$ which preserves orbits and the sense of direction in t. We say $T_\lambda(t)$ is *$\mathscr{A}$-stable at* λ_0 if there is a neighborhood V of λ_0 such that $T_\lambda(t) \sim T_{\lambda_0}(t)$ for all $\lambda \in V$.

Using results of Oliva or Morse-Smale systems (see Hale, Magalhães and Oliva [10]), one can prove

THEOREM 3.4. *Suppose* $T_\lambda(t)$ *satisfies the conditions of Theorem 3.2 and, in addition*,

(i) $T_\lambda(t)$ *is one-to-one on* $\mathscr{A}_\lambda$ *and the derivative* $DT_\lambda(t)$ *is an isomorphism for each* $\lambda \in \Lambda$, $\psi \in \mathscr{A}_\lambda$.

(ii) E_{λ_0} *is hyperbolic and* $W^u_{\lambda_0}(\phi) \pitchfork W^s_{\lambda_0,loc}(\psi)$ *for each* $\phi, \psi \in E_{\lambda_0}$ ($\pitchfork$ means transversal).

Then $T_{\lambda_0}(t)$ *is* $\mathscr{A}$*-stable*.

4. SOME EXAMPLES OF GRADIENT SYSTEMS. *4.1. A parabolic equation*. The simplest type of infintie dimension gradient system is a scalar parabolic equation

$$\begin{aligned} u_t &= u_{xx} + f(u) \quad \text{in } \Omega = (0,\pi) \\ u &= 0 \quad \text{in } \partial\Omega \end{aligned} \tag{4.1}$$

where $f \in C^2(R,R)$ satisfies

(4.2) $$\overline{\lim}_{|u|\to\infty} f(u)/u \le 0$$

The problem (4.1) generates a C^1-semigroup $T(t)$, $t \ge 0$ on $H_0^1(0,\pi)$ with $T(t)\phi(x) = u(t,x,\phi)$ being the solution of (4.1) with initial value ϕ at $t = 0$. The Liapunov function

$$V(\phi) = \int_0^{\pi} [\tfrac{1}{2}\,\phi_x^2(x) - F(\phi(x))]dx, \quad F(u) = \int_0^u f(s)ds$$

satisfies $dV(u(t,\cdot,\phi))/dt = -\int_0^1 u_t^2(t,x,\phi)dx \le 0$. This can be used to show that (4.1), (4.2) is gradient and orbits of bounded sets are bounded (see, for example, Henry [14] or Hale [6]). Therefore, if the equilibrium set E; that is, the solutions of

$$u_{xx} + f(u) = 0 \quad 0 < x < \pi$$

$$u = 0 \text{ at } 0,\pi$$

are bounded, then Eq. (4.1) has a compact attractor. The set E will certainly be bounded if $F(u) \to -\infty$ as $|u| \to \infty$.

An equilibrium point ϕ is hyperbolic if no eigenvalue of the operator $\partial^2/\partial x^2 + f'(\phi(\cdot))$ on $H_0^1(0,\pi)$ is zero. If all equilibrium points are hyperbolic, the attractor A is given by $A = \cup_{\phi\in E} W^u(\phi)$. Two rather remarkable results have recently been obtained concerning the flow on the attractor by Henry [15] (for other proofs, see Angenent [1], Hale [6]).

<u>THEOREM 4.1</u>. Suppose ϕ,ψ are hyperbolic equilibrium points. Then

(1) if there is an orbit of (4.1) with α-limit set ϕ and ω-limit set ψ, then $\dim W^u(\phi) > \dim W^u(\psi)$.

(2) $W^u(\phi)$ is transversal to $W^s(\psi)$.

For eq. (4.1), Theorem 4.1 implies the system is Morse-Smale if and only if the equilibrium points are hyperbolic. Even though (4.1) is an infinite dimensional problem, its dynamics is similar to a scalar ode in the sense that only local bifurcations occur. Henry [15] also shows that property 1) is true even if only one of the equilibrium points is hyperbolic. This latter result can be used to show how the equilibrium points on the attractor are connected by orbits (see also, Brunovsky and Fiedler [4]).

The proof of Theorem 4.1 consists in verifying the following facts:

(i) If a solution approaches ϕ as $t \to -\infty$ ($t \to +\infty$), it must have at most $\dim W^u(\phi) - 1$ zeros (at least $\dim W^u(\phi)$ zeros).

(ii) The number of zeros of solutions of a linear parabolic equation in one space dimension is nonincreasing in t.

Property (ii) can be found in Matano [21]. Property (i) is obtained by using stable and unstable manifold theory and connecting the tangent spaces with the eigenfunctions of the linear variational equation. The Sturmian theory relates the eigenfunctions to the zeros.

The next step would be to consider

$$(4.3)\qquad \begin{aligned} u_t &= \Delta u + f(u) \quad \text{in } \Omega \\ u &= 0 \quad \text{in } \partial\Omega \end{aligned}$$

where Ω is a smooth bounded domain in R^n. This system is also gradient under suitable conditions on f.

PROBLEM. Is some analogue of Theorem 4.1 true? A basic ingredient to the discussion of this problem is to determine a measure of complexity of the solution (in the one dimensional case, it was the number of zeros), relate this measure to the nth eigenvalue and show that complexity is nonincreasing along solutions. If Ω is a ball and one restricts the discussion to radially symmetric solutions, then the same theory seems to go through. However, without such assumptions, nothing seems to be known.

As remarked earlier, (4.1) is Morse-Smale if and only if equilibrium points are hyperbolic. Brunovsky and Chow [3] (see also Smoller and Wasserman [25]) have shown hyperbolicity is generic in f. Rocha [22] has given the same result with variable diffusion.

PROBLEM. For equation (4.3), are equilibrium points hyperbolic generically in (Ω,f)? The qualitative discussion of the flow for systems of parabolic equations is still in its infancy. It is extremely difficult to obtain effective methods for determining global existence of solutions and a priori bounds on the solutions. The maximum principle is a good tool for a scalar equation but is not true generally for systems. Attempts to extend the maximum principle by using invariant regions can be found for example in Smoller [24], but is very restrictive in applications. The method can sometimes be applied to systems which preserve cones (see, for example, Hirsch [16]). Other methods can be found in Rothe [23]. Much more research is needed in this area.

4.2. A damped wave equation. Consider the equation

$$(4.4)\qquad \begin{aligned} u_{tt} + 2\alpha u_t - \Delta u &= -f(u) - g(x) \quad \text{in } \Omega \\ u &= 0 \quad \text{in } \partial\Omega \end{aligned}$$

where $\alpha > 0$ is a constant, f is a C^2-function, $g \in L^2(\Omega)$, and Ω is a smooth bounded domain in R^n. The initial value problem for (4.3) is $u = \phi$, $u_t = \psi$ at $t = 0$. If $(\phi,\psi) \in H_0^1(\Omega) \times L^2(\Omega) \overset{\text{def}}{=} X$ and f satisfies certain growth conditions is negative enough for large u (see Babin and Vishik [2] or Hale [6]), then the function

$$V(\phi,\psi) = \int_\Omega [\tfrac{1}{2} |\nabla\phi(x)|^2 + \tfrac{1}{2} \psi^2(x) + F(\phi(x)) + g(x)\phi(x)]dx$$

is a Liapunov function, the equilibrium set E is bounded and orbits of bounded sets are bounded.

Furthermore, it is shown in [6] that $\alpha > 0$ implies the semigroup $T(t)$ generated by (4.3) satisfies

$$T(t) = S(t) + U(t)$$

where $U(t)$ is compact and $S(t)$ is linear with norm $\leq \exp(-\delta t)$ where $\delta > 0$ is a constant. Thus, (4.3) has a compact attractor A. From the definition of a compact attractor, this implies that $\omega(B)$ for any bounded set $B \subset X$ belongs to A. Haraux [13] has obtained the same result using different methods. Babin and Vishik [2] obtained a weaker conclusion concerning the attractivity properties of A.

If the equlibrium points are hyperbolic, then $A = \cup_{\phi \in E} W^u(\phi)$. From the results in Hale and Scheurle [12], it follows that, for any $\psi \in A$, the map $t \to T(t)\psi$ is as smooth as f; a rather surprising phenomena since $T(t)\eta$ is not differentiable in t for every η.

If we consider the attractor A as a function of α, $A = A_\alpha$, then it is natural to discuss the dependence of A_α on α. The index $i(\phi) = \dim W^u_\alpha(\phi)$ does not depend on α. Thus, hyperbolicity is true for all α if it is true for one α.

PROBLEM. *Is $W^u_\alpha(\phi)$ transversal to $W^s_\alpha(\psi)$ for all equilibrium points* ϕ, ψ? Preliminary results of Hale, Lin and Rocha indicate that for α large this is true.

For systems of hyperbolic equations

$$u_{tt} + 2Bu_t - D\Delta u = -f(u) - g(x)$$

where $u \in R^N$ and B,D are diagonal positive matrices, the semigroup if it exists satisfies the property mentioned above. Thus, if it is point dissipative and orbits of bounded sets are bounded, then the compact attractor exists. Of course, showing the global existence, etc. involves the same problems mentioned before for parabolic systems.

4.3. *Scalar retarded functional differential equations*. Consider the scalar equation

$$\dot{x}(t) = -\int_{-1}^{0} b(\theta)g(x(t+\theta))d\theta \tag{4.5}$$

where g is a C^1-function, $\int_0^x g(s)ds \to \infty$ as $|x| \to \infty$, where b is a C^2-function with $b(\theta) > 0$, $b'(\theta) \geq 0$, $b''(\theta) \geq 0$ and there is a $\theta_0 \in [-1,0]$ such that $b''(\theta_0) > 0$. If $C = C([-1,0],R)$ and $x(t,\phi)$ is the solution of (4.4) with initial value $\phi \in C$, then $T(t)$: $C \to C$, $t \geq 0$, defined by $[T(t)\phi](\theta) = x(t+\theta),\phi)$, $-1 \leq \theta \leq 0$, is a C^1-semigroup. It is also gradient (see, for example, Hale and Rybakowski [11] or Hale [6]). Furthermore, the set of equilibrium points E are constant functions corresponding to the zeros of g. If $\phi \in E$ is hyperbolic, then $\dim W^u(\phi) \leq 1$. Since $T(t)$ is compact for $t \geq 1$, it follows that there is a compact attractor A if E is bounded. Also, $A = \cup_{\phi \in E} W^u(\phi)$ if each $\phi \in E$ is hyperbolic. Thus, $\dim A = 1$. It is shown in Hale and Rybakowski [11] that there are examples g for which $W^u(\phi)$ is not transversal to $W^s(\psi)$ in contrast to the parabolic equation in (4.1). There is also another interesting property of this system. It is gradient, but there are complex eigenvalues corresponding to equilibria. The unstable manifold always corresponds to a real eigenvalue.

<u>PROBLEM</u>. <u>For general gradient systems, does</u> $W^u(\phi)$ <u>correspond to real eigenvalues?</u>

The above example showsthat scalar gradient delay equations are more complicated dynamically than a parabolic equation. However, the fact that it is a scalar does put restricitions on the flow. It is an interesting problem to discuss the nature of these restrictions in a more general setting.

We only mention two results. The first one is that the center manifold equations for general scalar retarded functional differnetial equations $\dot{x}(t) = f(x_t)$, $x_t(\theta) = x(t+\theta)$, $-1 \leq \theta \leq 0$, must always correspond to an mth order ode and not a general system of m ode's (see Hale [7]).

The other example is due to Mallet-Paret [18]. For the equation

$$\dot{x}(t) = -\beta x(t) - g(x(t-1))$$

with $yg(y) > 0$ and $g'(0) > 0$ (negative feedback), he showed there was a Morse decomposition (in the sense of Conley) for the flow. The Liapunov function essentially corresponds to the number of zeros of a solution in an interval $(t-1,t]$.

BIBLIOGRAPHY

1. S. B. Angenent, "The Morse-Smale property for a semilinear parabolic equation", Report No. 11, Math. Inst. Univ. Leiden, June, 1984, to appear in J. Differential Equations.

2. A. V. Babin and M. I. Vishik, "Regular attractors of semigroups of evolutionary equations", J. Math. Pures et Appl. 62 (1983), 441-491.

3. P. Brunovsky and S.-N. Chow, "Generic properties of stationary states of reaction diffusion equations", J. Differential Equations 53 (1984), 1-24.

4. P. Brunovsky and B. Fiedler, "Heteroclinic connections of stationary solutions of scalar reaction diffusion equations", Proc. PDE-Semester, Stefan Banach Institute, 1984, B. Bojarski (Ed.), Warsaw, Poland.

5. G. D. Cooperman, "α-condensing maps and dissipative systems", Ph.D. Thesis, Brown University, Providence, RI, 1978.

6. J. K. Hale, "Asymptotic behavior and dynamics in infinite dimensions", LCDS Report #84-28, Div. of Appl. Math., Brown Univ.; to appear in Pittman Research Notes, Editors, Hale and Martinez-Amores, October, 1985.

7. J. K. Hale,"Flows on center manifolds for scalar functional differential equations", Proc. Royal Society of Edinburgh, to appear in 1985.

8. J. K. Hale, J. P. LaSalle, and M. Slemrod, "Theory of a general class of dissipative processes", J. Math. Ana. Appl. 39 (1972), 177-191.

9. J. K. Hale and O. Lopes, "Fixed point theorems and dissipative processes", J. Differential Equations 13 (1973), 391-402.

10. J. K. Hale, L. T. Magalhães and W. M. Oliva, An Introduction to Infinite Dimensional Dynamical Systems - Geometric Theory, Appl. Math. Sci., Vol. 47, Springer-Verlag.

11. J. K. Hale and K. Rybakowski, "On a gradient-like integro-differential equation", Proc. Royal Society of Edinburgh, 92A (1982), 77-85.

12. J. K. Hale and J. Scheurle, "Smoothness of bounded solutions of nonlinear evolution equations", J. Differential Equations 56 (1985), 142-163.

13. A. Haraux, "Some remarks on dissipative hyperbolic problems", preprint of a lecture in College de France, May 1984.

14. D. Henry, "Geometric Theory of Semilinear Parabolic Equations", Lecture Notes in Mathematics No. 840, Springer-Verlag, Berlin/Heidelberg/New York, 1981.

15. D. Henry, "Some infinite dimensional Morse-Smale systems defined by parabolic partial differential equations", preprint, to appear in J. Differential Equations.

16. M. W. Hirsch, "Stability and convergence in strongly monotone maps", preprint, May 1984.

17. J. Mallet-Paret, "Negatively invariant sets of compact maps and an extension of a theorem of Cartwright", J. Differential Equations 22 (1976), 331-348.

18. J. Mallet-Paret, Morse decomposition and global continuation of periodic solutions of singularly perturbed delay equations", in Systems of Nonlinear Partial Differential Equations (Ed. J. Ball), Reidel, 1983.

19. R. Mañé, "On the dimension of the compact invariant set of certain nonlinear maps", Lect. Notes in Math. 898 (1981), Springer-Verlag, 230-242.

20. P. Massatt, "Attractivity properties of α-contractions", J. Diff. Equations 48 (1983), 326-333.

21. H. Matano, "Nonincrease of lap number of a solution for a one-dimensional semilinear parabolic equation", Pub. Fac. Sci. Univ. Tokyo, Sec. 1A, 29 (1982), 401-441.

22. C. Rocha, "Generic properties of equilbria of reaction-diffusion equations with variable diffusion", Proc. Royal Soceity of Edinburgh, 1985.

23. F. Rothe, "Global solutions of reaction-diffusion systems", Lecture Notes in Math. 1072 (1984), Springer-Verlag.

24. J. Smoller, "Shock waves and reaction-diffusion equations", Grund. Math. Wissenshaften 258, Springer-Verlag, 1982.

25. J. Smoller and A. Wasserman, "Generic bifurcation of steady state solutions", J. Diff. Equations 52 (1984), 432-438.

DIVISION OF APPLIED MATHEMATICS
BROWN UNIVERSITY
PROVIDENCE, RI 02912

Contemporary Mathematics
Volume **58**, Part III, 1987

E. CARTAN'S GENERALIZED SPACES AND CONNECTIONS

MASATAKE KURANISHI

A systematic way of investigating a geometry of a structure is by means of constructing a Cartan connection. This has been accomplished, for example, in the cases of Riemann geometry, conformal geometry, and CR structures with non-degenerate Levi-forms. E. Cartan in his Notice sur les travaus scientifiques [2]developed the notion of the generalized spaces and indicated how to study the geometry of such spaces by Cartan connections. C. Ehresmann in [4] made the notion more precise. In this article we develop this approach further.

§1 PRE-GENERALIZED SPACES

We start with a Lie group G acting transitively, effectively, and smoothly on a manifold X. We also fix a reference point P_0 in X. Thus we may identify X with G/H where H is the isotropy group of P_0. The idea is to consider spaces which look locally at each point like the homogenous space $X = G/H$. We thus start by attaching X to each point of a smooth manifold M. Namely, for each point p in M we consider a set $A(p)$ of germs of diffeomorphisms of open neighborhoods of P_0 onto neighborhoods of p sending P_0 to p. We consider elements in $A(p)$ up to an equivalence. This notion of equivalence should satisfy the following:

(*) If f is in $A(p)$ and $h \in H$, $f \circ h$ is also in $A(p)$. If f_1 and f_2 are in $A(p)$, there is a unique h in H such that f_1 and $f_2 \circ h$ are equivalent.

In the above h is regarded as a map $(X, P_0) \to (X, P_0)$ by its action on X. Any $A(p)$ as above may be called a set of

Partially supported by the National Science Foundation under grant No. NSF DMS 84-02710.

© 1987 American Mathematical Society
0271-4132/87 $1.00 + $.25 per page

attaching maps at p. An equivalence class in A(p) is called a frame and the set denoted by F_p. The operation of H (from the right) on A(p) induces an operation of H from the right on F_p. The condition (*) implies that the operation of H on F_p is simply transitive. Assume that we have a set A(p) of attaching maps at each point p in M. Then we have the space of frames F_p for each p. We say that we have a pre-generalized space patterned after (G,X) when $F = \bigcup\{F_p;\ p \in M\}$ forms canonically a smooth principal bundle over M with the structure group H where the operation of H is as indicated above. Here the smoothness is the one coming from smooth families of germs of diffeomorphisms. F is called the principal bundle of frames of the pre-generalized space M. We thus have

(1) $$\rho : F \to M$$

In all the known examples the attaching maps are obtained by using the spaces of jets. Namely, for an integer ℓ denote by $J^\ell(p)$ the space of ℓ-jets of local diffeomorphism of (X, P_0) to (M,p). Clearly the action of H on X induces its action on $J^\ell(p)$. Let $\tilde{J}^\ell(p)$ be a submanifold of $J^\ell(p)$ preserved under the action of H. We construct a manifold $K^\ell(p)$ on which H acts without fix points. We then consider a smooth map R $\tilde{J}^\ell(p) \to K^\ell(p)$ which commutes with the action of H. Finally we pick an orbit B(p) of H in $K^\ell(p)$. We call f an attaching map at p when ℓ-jet at P_0 is in $\tilde{J}^\ell(p)$ and the image by R is in B(p). Attaching maps are equivalent when their image by R of their ℓ-jets at P_0 are the same. Since H acts without fixed points in $K^\ell(p)$, we see clearly that A(p) thus obtained satisfies our condition (*).

In the case of Riemann geometry, $X = \underline{R}^n$, P_0 = the origin, and G is the group of euclidean motions. We take $\ell = 1$ and let $\tilde{J}^1(p) \subset J^1(p)$ be the set of 1-jets of f such that df at P_0 is the isometry of the tangent spaces. $K^1(p) = \tilde{J}^1(p)$ with R the identity map. The orbit B(p) is equal to $K^1(p)$. In the case of the conformal structure, we also consider $\underline{R}^n$ with the conformal structure induced by its standard metric. However, to make it a homogeneous space we have to add points at infinity. There is a classical way to do this. We thus obtain a manifold $X \supset \underline{R}^n$ with a conformal structure (extending the standard one in $\underline{R}^n$) on which the group of conformal transformations acts transitively. In this case ℓ is 2. $\tilde{J}^2(p)$ is the set of 2-jets of f such that the induced conformal structure and the standard conformal structure at P_0 agrees up to order one. It is not obvious but we can show that the isotropy group H acts simply transitively on $\tilde{J}^2(p)$ provided $n \geqslant 3$. Thus we can associate a unique pre-generalized space (patterned after G/H) for any conformal structure of dimension $n \geqslant 3$. Since all non-trivial holomorphic maps are conformal with respect to the standard metric of $\underline{C}$, we need an infinite group rather than a Lie group to handle the case of $n = 2$.

For a CR structure we take $\ell = 2$. However, $\tilde{J}^\ell(p)$ and $K^\ell(p)$ are no longer the same. A CR structure on a manifold M is a subbundle T'' of the complex tangent vector bundle $\underline{C}\,TM$ satisfying the following conditions : 1) If X and Y are sections of T'', so is $[X,Y]$. 2) $T'' \cap T' = \{0\}$ where T' is complex conjugate (or the bar) of T''. 3)

(1) $$S_T'' = \underline{C}TM/(T'' + T')$$

has fiber complex dimension 1. For a CR structure we have the Levi-form: For sections X and Y of T''.

(3) $$C_{T''}(X,Y) = -i[X,Y] \mod T' + T''.$$

It is a hermitian form valued in $S_{T''}$. Note that $S_{T''}$ has the bar of $\underline{C}TM$. If M is a real hypersurface of codimension one in a complex manifold, the set of complex tangent vectors to M, which are of type $(0,1)$ when expressed in terms of a complex analytic chart of the ambient complex manifod, is a CR structure. In particular, for a domain U with a smooth boundary in a complex manifold, its boundary M has a CR structure. The main interest of CR structures comes from its relation to the function theory in U.

The geometry of CR structures with non-degenerate Levi-form was constructed first by E. Cartan [1] in the case of dimension 3. The construction was generalized by Tanaka [5],[6] and by Chern-Moser [3] for general dimensions. The construction of the frame bundle in the above works is different from the one given by means of attaching maps outlined above. To construct the frame bundle from the view point of pre-generalized space, we must first construct the homogeneous space $X = G/H$. This is the homogeneous space constructed in [3] and also in [6]. Note that we have a Levi-form at each point of a CR structure M. However, when M is connected, they are equivalent. Thus we have a unique equivalence class of non-degenerate hermitian forms attached to M. Let X be the homogeneous CR structure with the same equivalence class of Levi-forms as that of M. Denote by $T''_X \subset \underline{C}TX$ the CR structure of X. Rather than formally introducing $\tilde{\mathcal{J}}^2(p)$, $K(p)$, and R, we shall state directly a definition of attaching maps. Let $f\colon (X',P_0) \to (M',p)$ be a diffeomorphism (where X' (resp. M') is an open neighborhood of P_0 (resp. p). f is an attaching map if 1) df sends T''_X at P_0 to T'' at p and 2) The pull back by f of the dual space of $S_{T''}$ and the dual space of $S_{T''_X}$, both considered as a subspace of 1-forms, have contact of order 1 at P_0. Attaching maps f_1 and f_2 are considered equivalent when their differentials agree at P_0 and $(f_1^{-1})^*\,\theta_X \equiv (f_2^{-1})^*\,\theta_X \mod m_p^2$, where θ_X is a generator of the dual space of $S_{T''_X}$ and m_p is the maximal ideal in the algebra of the germs at p of smooth functions on M. Since we know H explicitly (cf. [3]) we can check that our definition of equivalence satisfies the condition (*). It is then obvious that our attaching maps are constructed by means of 2-jets as described above.

§2 GENERALIZED SPACES

In a pre-generalized space we have attaching maps and frames at each point. However, there is no guideline to relate these objects at infinitesmally near points. In order to examine the situation we first consider the case of homogeneous spaces.

Let X, P_0, G, and H be as in the beginning of §1. We note that we have a canonical pre-generalized structure on X. Namely, for P in X pick g in G which sends P_0 to P. The operation of g on X is a diffeomorphism $f_g\colon (X,P_0) \to (X,P)$. The set of germs at P_0 induced by such f_g (with the trivial equivalence relation) obviously satisfies our condition (*). Clearly, the manifold F_p of frames can be identified with $G_P = \{g \in G;\ f_g(P_0) = P\}$. Thus the principal bundle F_X of frames may be identified with

(4) $$F_X = G \to G/H = X$$

In this case X is homogeneous and hence we may translate frames: The frame represented by $f_{g'} \mapsto$ the frame represented by $f_{gg'}$

for a fixed g in G. One way of programming this process is by means of the Mauer-Cartan forms. Namely, we pick a base w_X^j of left invariant Pfaffian forms on $G = F_X$. Denote $\bar{F}_X$ (resp. by W_X^i) a copy of F_X (resp. of w_X^i). Then we recover the above translations by solving the Pfaffian systems on $F_X \times \bar{F}_X$.

$$W_X^j - w_X^j = 0. \tag{5}$$

Thus F_X together with w_X^i contain enough information to construct the homogeneous space structure of X. In order to write more concisely we denote by $\underline{G}$ the Lie algebra of G and consider $\{w_X^i\}$ as a $\underline{G}$-valued Pfaffian form

$$TF_X \to \underline{G} \tag{6}$$

There are properties satisfied by the Mauer-Cartan form (6) which may be described by using only the canonical pre-generalized space X. Now we ask for a pre-generalized space M a Pfaffian form

$$w\colon TF \to \underline{G} \tag{7}$$

having these properties. We thus arrive at the notion of Cartan connections (and also associated Cartan connections) for a pre-generalized space. The conditions are the following: I) For each $\underline{f}$ in F the restriction of w to the tangent space $T_f F$ to F at $\underline{f}$ is an isomorphism of vector spaces. II) If $R_h\colon F \to F$ is the right operation of h in H

$$R_h^* w = \mathrm{Ad}(h^{-1})w \tag{8}$$

where $\mathrm{Ad}(h^{-1})$ is the automorphism of $\underline{G}$ induced by the automorphism $g \mapsto h^{-1}g\,h$ of G. (III) Denote by $\underline{h}(t)$ the one-parameter subgroup of H generated by $\underline{h}$ in the Lie algebra $\underline{H}$ of H and let $X_{\underline{f}}$ be the tangent vector at $t = 0$ of the curve $t \mapsto R_{\underline{h}(t)}(\underline{f})$. Then $w(X_{\underline{f}}) = \underline{h}$.

When these conditions are satisfied, we say that w is a Cartan connection on F (with respect to $G/H = X$). However, under these conditions there are not enough relations between the attaching maps and the Cartan connections. Thus we put the following conditions: (IV) For any frame $\underline{f}$ at p represented by an attaching map f we have the commutative diagram

$$\text{(9)} \qquad \begin{array}{ccc} w:\ T_{\underline{f}}F & \longrightarrow & \underline{G} \\ \downarrow & & \downarrow \\ df^{-1}:\ T_pM & \longrightarrow & \underline{G}/\underline{H} = T_I(G/H) = T_{P_0}(X) \end{array}$$

where I denotes the identity element of G modulo H, and where the vertical arrows are the canonical projections. A Cartan connection satisfying the condition (IV) will be called an associated Cartan connection for the pre-generalized space.

It is easy to check that these conditions are satisfied by the Mauer-Cartan form (6). It is also easy to see that an associated Cartan connection for a pre-generalized space exists locally, provided df^{-1} at p is uniquely determined by its equivalence class. Hence the following condition is the necessary condition for an existence of an associated Cartan connection:

(**) If f_1 and f_2 in $A(p)$ are equivalent, the 1-jets, of f_1 and f_2 at P_0 are equal.

By a generalized space we mean a pre-generalized space together with a choice of an associated Cartan connection.

The curvature of a Cartan connection w is given by

$$\text{(10)} \qquad K = dw + \tfrac{1}{2}[w,w]$$

where $[w,w]$ is defined in terms of the Lie bracket in $\underline{G}$. In the case of the Mauer-Cartan form w_X the curvature is zero. Thus the curvature measures the deviation of a generalized space from the homogeneous space $X = G/H$. For each frame $\underline{f}$ at p there is $\underline{H}$-valued 2-form $K_{\underline{f}}$ on T_pM such that K evalued at $\underline{f}$ is equal to $\rho^*K_{\underline{f}}$ where ρ is the projection $F \to M$. It is easy to see that for $h \in H$

$$\text{(11)} \qquad K_{R(h)\underline{f}} = \mathrm{Ad}(h^{-1})\, K_{\underline{f}}$$

A choice of an associated Cartan connection for a given pre-generalized space is not unique. If we use a local trivialization $F|M' \approx M' \times H$ for a small open submanifold $M' \subset M$, a general associated Cartan connection w over M' is given by

$$w = \mathrm{Ad}(h^{-1})\upsilon + w_H \tag{12}$$

where υ is an arbitrary $\underline{G}$ valued Pfaffian form on M' and w_H is the Mauer-Cartan form on H. υ mod $\underline{H}$ is determined by the commutativity of (9): for p in M'

$$(df^{-1})_p \equiv (\upsilon)_p \qquad \mathrm{mod}\ \underline{H} \otimes T^*_pM \tag{13}$$

where f represents the frame which corresponds to (p, I_H) in the local chart (I_H being the identity element of H). Thus the set of associated Cartan connections is parametrized by $\underline{H}$-valued Pfaffian forms.

The above observation suggests that it is desirable to put conditions on associated Cartan connections to reduce the number of parameters. If we can find global conditions to make the choice of associated Cartan connections unique, it will also help to define the generalized space associated with a pre-generalized space, because, among others, it makes the definition of w global (in view of the fact that the existence of an associated Cartan connection is proved only locally). The standard way of doing this is to put conditions on the curvature.

Denote by $\rho_{\underline{G}/\underline{H}}$ the projection $\underline{G} \to \underline{G}/\underline{H}$. $\rho_{\underline{G}/\underline{H}} \cdot K$ is called the torsion of the connection. In the case of three examples mentioned in §1, we can always find an associated Cartan connection with zero torsion. Let us find conditions for local existence of associated Cartan connections without torsion for a pre-generalized space patterned after $X \doteq G/H$. The problem being local, we may assume that F has a trivialization $M \times H$. Also we fix a vector space decomposition

$$\underline{G} = \underline{H} + \underline{M} \tag{14}$$

so that $\underline{G}/\underline{H}$ may be identified with $\underline{M}$. Denote by $\underline{f}_p$ the frame corresponding to $p \times I_H$ in the trivialization. For each $\underline{m} \in \underline{M}$ there is a unique vector field $X_{\underline{m}}$ to M so that

$$w((X_{\underline{m}})_p) \equiv \underline{m} \qquad \mathrm{mod}\ \underline{H} \tag{15}$$

for all p in M, where we made an identification $T_pM \subseteq T_{\underline{f}_p}F$ in terms of the trivialization $F = M \times H$. By the commutativity of (9), the vector field $X_{\underline{m}}$ is uniquely determined by pre-generalized space structure and the trivialization. For a general w we then have

$$(16)\qquad w((X_{\underline{m}})_p) = \underline{m} + u((X_{\underline{m}})_p)$$

where u is an arbitrary $\underline{H}$-valued Pfaffian form on M. We wish to find u so that the torsion is zero. Since

$$dw(X,Y) = X\,w(Y) - Y\,w(X) - w([X,Y])$$

for tangent vector fields X,Y we see by (10) that the condition is

$$(17)\qquad w([X_{\underline{m}},X_{\underline{m}'}]) - [\underline{m},\underline{m}'] \equiv [\underline{m},u(X_{\underline{m}'})] - [m',u(X_{\underline{m}'})] \quad (\text{mod } \underline{H}).$$

As noted above the left side of the above mod $\underline{H}$ is determined by the pre-generalized space structure, provided we fix a trivialization of M and the decomposition (14). Hence (17) is an equation for an unknown $\underline{H}$-valued Pfaffian form u on M. Since $X_{\underline{m}}$ is determined by $\underline{m}$ at each point, we may consider $u(X_{\underline{m}})$ as a linear map of $\underline{G}/\underline{H}$ at each p, say $\underline{u}(\underline{m})$. The formula suggests we introduce

$$(18)\qquad \partial\colon \Lambda^1(\underline{G}/\underline{H}) \otimes \underline{H} \to \Lambda^2(\underline{G}/\underline{H}) \otimes \underline{G}/\underline{H}$$

by

$$\partial\underline{u}(\underline{m},\underline{m}') = P_{\underline{G}/\underline{H}}([\underline{m},\underline{u}(\underline{m}')] - [\underline{m}',u(\underline{m})])$$

and set $C_p \in \Lambda^2(\underline{G}/\underline{H}) \otimes \underline{G}/\underline{H}$ by

$$C_p(\underline{m},\underline{m}') = P_{\underline{G}/\underline{H}}((df^{-1})_p([X_{\underline{m}},X_{\underline{m}'}]) - [\underline{m},\underline{m}'])$$

where f is a representative of $\underline{f}_p$. Thus condition for an existence of an associated Cartan connection without torsion is given by the formula

$$(19)\qquad C_p \in \text{Image}\,\partial \qquad \text{for all } p \in M$$

This condition is, by the nature of the construction, obviously independent of the trivialization and the decomposition (14) and determined by the pre-generalized space.

In the case of Riemann geometry, an associated Cartan connection without torsion exists and is unique. We thus obtain the Levi-Civita connections. In the cases of two other examples in §1, the condition (19) is also satisfied. However, we can impose a stronger condition:

$$K \equiv 0 \qquad \operatorname{mod}[\underline{H}, \underline{H}] \; . \tag{20}$$

We call such an associated Cartan connection admissible. We can analize the admissibility of an associated Cartan connection by exactly the same method we used for torsion. However, even in the cases of two examples mentioned above, an admissible associated Cartan connections is not unique. We have to impose further conditions on the curvature. However, the general nature of this procedure seems to be not yet well understood.

REFERENCES

[1] Cartan, E., Sur la geometrie pseudo-conform des hypersurfaces de deux variables complexes, I. Ann. Math. pura Appl.,(4) 11(1932) 17-90; II Ann. Scuola Norm. Sup. Pisa,(2)1(1932) 33-354.

[2] Cartan, E. Notice sur les Travaux Scientifiques, Oeuvres I, 1. 1-105.

[3] Chern, S.S. and Moser, J., Real hypersurfaces in complex manifolds, Acta Math. 133(1974) 219-271.

[4] Ehresmann, C., Les connexions infinitesimales dans un espace fibré differentiable. Colloque de Topologie, Bruxelles (1950), 29-55.

[5] Tanaka, N. Graded Lie algebras and geometric structures, Proc. US-Japan Seminar in Differential Geometry, 1965, 147-150.

[6] Tanaka, N., On non-degenerate real hypersurfaces, graded Lie algebras and Cartan connections. Japan J. Math. 2(1976) 131-190.

[7] Tanaka, N., On the equivalence problems associated with simple graded Lie algebras. Hokkaido Math. Jour. 8(1979) 23-84.

Department of Mathematics
Columbia University
New York, NY 10027

Contemporary Mathematics
Volume 58, Part III, 1987

INFINITY MANIFOLDS FOR POSITIVE ENERGY IN CELESTIAL MECHANICS

Ernesto A. Lacomba[1]

ABSTRACT. We describe here the asymptotic behavior of escape orbits for the n-body problem of celestial mechanics when the energy is positive. This is made by means of blowing up the infinity of configuration space. This way one makes precise how the orbits defined for all time t, must escape as $t \to +\infty$. In fact, a neighborhood of infinity in the energy surface is foliated by submanifolds of orbits escaping to infinity in any given direction of configuration space. This confirms the classical result that positive energy makes escape of particles easy.

1. INTRODUCTION. In this paper we describe how positive energy orbits in celestial mechanics are escape orbits as $t \to \pm\infty$, unless they are stopped on either side by a singularity which is not a binary collision. The main tool is a resolution or blow up at infinity of configuration space, which was first defined by this author and Simo |4|.

We begin by showing that there is no loss of generality in assuming that $h = +1$, by taking advantage of homogeneities of the potential and kinetic energy. In the same way, one may assume $h = -1$ when studying negative energy surfaces. This is related to Kepler's third law.

It is made precise how a neighborhood of infinity is foliated by submanifolds consisting of orbits escaping to infinity in any given direction of configuration space. Symmetrically, there is another neighborhood of infinity foliated by orbits coming from infinity in any given direction. The reason is that the infinity manifold has two normally hyperbolic submanifolds of equilibrium orbits: one of them attracts all orbits escaping to infinity, while the other one repels all orbits coming from infinity. Some partial results in this direction were announced in |3|.

1980 Mathematics subject classification 58F05, 58F15, 70F10

[1]Partially supported by CONACYT grant PCCBBNA 022553 (México)

© 1987 American Mathematical Society
0271-4132/87 $1.00 + $.25 per page

2. IMPORTANCE OF ENERGY SIGN AND HOMOGENEITY.- In general an n-body problem can be stated in the form

$$(1) \qquad \begin{aligned} \dot{q} &= pA^{-1} \\ \dot{p} &= \operatorname{grad} U(q) \end{aligned}$$

where $p \in \mathbb{R}$; $q \in N \subset \mathbb{R}^k$, the set N being an open cone, and $U: N \to \mathbb{R}^+$ (<u>potential energy</u>) is analytic homogeneous of degree -1. The k x k matrix A is positive definite. The Hamiltonian or <u>total energy</u> is $H(q,p) = K(p) - U(q)$ where

$$K(p) = 1/2\ pA^{-1}p^t$$

is the kinetic energy. Given $h \in \mathbb{R}$, we define the corresponding energy level by $E_h = \{(q,p) \in N \times \mathbb{R}^k : H(q,p) = h\}$. The equation defining it or <u>energy relation</u> is written as

$$(2) \qquad K(p) = U(q) + h$$

The projection of E_h to configuration space is the so-called <u>Hill region</u> $U + h \geq 0$. This is a manifold with boundary $U = -h$ (the zero velocity points), if $h < 0$. If $h \geq 0$ the boundary is empty, and there is no restriction.

If $h < 0$, in some sense U as attracting potential predominates over K, energy due to the motion (see (2)), giving the possibility of bounded or recurrent motions.

If on the contrary $h > 0$, then K predominates over U making it easier to escape, unless the motion is abruptly stopped by a non regularizable singularity. We are going to make precise how this occurs.

We end up this section by showing that because of the homogeneity of U and K there are only 3 different phase portraits of energy levels $h = -1$, $h = 0$ and $h = +1$ from the qualitative viewpoint. This evidence supports even more the fact that information about the phase portrait is simplified by looking at energy levels. More precisely,

PROPOSITION 1.- For each $\lambda > 0$, there is a homothetical diffeomorphism f_λ of phase space taking E_h onto $E_{\lambda h}$ for any energy $h \in \mathbb{R}$. A change of time scale takes solutions exactly into solutions.

<u>Proof</u>: We define the diffeomorphism by $f_\lambda(q,p) = (\lambda^{-1}q, \lambda^{1/2}p)$. Then we compute $H(f_\lambda(q,p)) = K(\lambda^{1/2}p) - U(\lambda^{-1}q) = \lambda H(q,p)$. So that

$(q,p) \in E_h$ implies $f_\lambda(q,p) \in E_{\lambda h}$. The change of time scale $t \to \lambda^{-3/2} t$ takes solutions into solutions, as we check in the system (1).

Q.E.D.

A particular case of this proposition was considered by Simó |8| for the isosceles 3-body problem. However, it does not seem to be previously known in its general case.

We remark that the same powers of λ involved in the change of q,p,t, appear for the radial coordinate r when we perform a blow up of the origin to study total collisions |6|. This is no surprise, since both transformations derive naturally from homogeneity considerations.

If we set $\mu = \lambda^{-1/2}$ in the above transformations, we get for position and time $q \to \mu^2 q$, $t \to \mu^3 t$, which is a generalized statement of Kepler's third law: a homothety multiplying positions by the square of some parameter and time by the cube of the same parameter, takes solutions onto solutions in the n-body problem. This automatically implies that velocities or momenta are multiplied by μ^{-1}.

3. BLOW UP AT INFINITY FOR POSITIVE ENERGY. We restrict ourselves from now on to the case $h > 0$, and we may without loss of generality assume $h = +1$, whenever necessary.

If we denote by $I = qAq^t$ the moment of inertia of the system, the Lagrange-Jacobi identity |7| states that $\ddot{I} = 2(2K-U) = 2(U+2h)$, that is $\ddot{I} \geq 4h$. Integrating twice for $h > 0$, we conclude that for any motion defined for all time $t \in \mathbb{R}$, $r(t) \geq \alpha|t|$, where $r = I^{1/2}$ is the radius of inertia and $\alpha > 0$. This means that I turns out to be a strictly convex function of t with a unique minima, and satisfying $I \to \infty$ as $t \to \pm\infty$. We will find below a similar property for r as function of a fictitious time τ.

We remark from (2) that because of the homogeneity $U(q) \to 0$ if $q \to \infty$ radially. This implies $K(p) \to h$, which can be interpreted as $|p|$ aproaching the "hyperbolic speed" $\sqrt{2h}$, from the definition of K. This becomes more evident in equation (4), below.

We defined the blow up at infinity for positive energy |4|, by the transformation

$$\rho = I^{-1/2}, \quad Q = \rho q, \quad dt = \rho^{-1} d\tau,$$

with momenta p being kept the same. The change to afictitious time scale τ speeds up solutions as $\rho \to 0$. It is needed in order to get rid of an annihilating factor in the differential equations for $\rho = 0$. The system (1) becomes

$$\begin{aligned} \rho' &= -\rho v \\ Q' &= pA^{-1} - vQ \qquad (3) \\ p' &= \rho \operatorname{grad} U(Q) \end{aligned}$$

where $v = pQ^t$, $Q \in S = \{q \in N : qAq^t = 1\}$ and the prime denotes derivative with respect to τ. The energy relation becomes

$$\tfrac{1}{2} pA^{-1}p^t = \rho U(Q) + h \qquad (4)$$

It was Mc Gehee |6|, who first defined a blow up to study total collisions on any energy surface.

Taking now ρ, v, Q, Q' as more convenient coordinates, we rewrite (3) and (4) as

$$\begin{aligned} \rho' &= -\rho v \\ v' &= Q'AQ'^t - \rho U \qquad (5) \\ Q'' &= \rho \operatorname{grad}_Q U - (Q'AQ'^t)Q - vQ', \end{aligned}$$

$$\tfrac{1}{2}(Q'AQ'^t + v^2) = \rho U + h, \qquad (6)$$

where $(Q,Q') \in TS = \{(Q,Q') : Q \in S,\ Q'AQ^t = 0\}$. The flow is still defined and non trivial for $\rho = 0$, which defines an invariant infinity manifold N_h, where

$$\bar{E}_h = E_h \cup N_h = \{(\rho,v,Q,Q') : \rho \geq 0, (Q,Q') \in TS \text{ and } (6) \text{ holds}\}.$$

From (5) and (6) we see that equilibrium points occur only for $\rho = 0$, and we have in fact a whole submanifold of equilibria

$$S_h = \{(\rho,v,Q,Q') : \rho = 0, Q \in S, Q' = 0,\ v^2 = 2h\} \subset N_h$$

This submanifold has 2 connected components $S_h^{\pm}$ corresponding to the 2 possibilities $v = \pm\sqrt{2h}$. In coordinates ρ, Q, p this condition is written as $p = \pm\sqrt{2h}\, QA$. A dimensional count shows that $\dim S_h = \tfrac{1}{2}\dim N_h$, and more precisely we have $S_h = S \times S^0$ and $N_h = S \times S^{k-1}$, topologically. The following result is a refinement of |3|.

THEOREM 1.- The vector field on $\bar{E}_h$ has S_h^+ as a normally hyperbolic attractor and S_h^- as a normally hyperbolic repeller. The (un) stable manifold of $S_h^{(\pm)}$ is an open neighborhood in $\bar{E}_h$ foliated by the k-dimensional (un) stable submanifolds of the individual equilibrium points.

Proof: This is a very special case of normal hyperbolicity, and we just have to check that for any $x_0 \in S_h$ there is a complement of the tangent space $T_{x_0}S_h$ in $T_{x_0}\bar{E}_h$ which is contracted (if $v_0 = \sqrt{2h}$) or stretched (if $v_0 = -\sqrt{2h}$) by the linear part of the flow in $\bar{E}_h$. If we write $x_0 = (0, v_0, Q_0, 0)$, a computation shows that the order 2k matrix B of the linear part of the system at x_0 is

$$B = \left[\begin{array}{cc|cc} -v_0 & 0 & 0 & 0 \\ -U & 0 & 0 & 0 \\ \hline 0 & 0 & 0 & I \\ (\mathrm{grad}_Q U)^t & 0 & 0 & -v_0 I \end{array}\right]$$

where the lower right block consists of square $k-1$ matrices, and I denotes the corresponding identity. By computing the gradient of the function in (6), we check that B leaves invariant the subspace $T_{x_0}\bar{E}_h$ in phase space.

Solving for v in the energy relation (6), we see that ρ, Q, Q' are good coordinates to parametrize $T_{x_0}\bar{E}_h$. A careful computation using the chain rule, shows that with this parametrization the order $2k-1$ matrix of the restriction of B to $T_{x_0}\bar{E}_h$ is obtained from B above by deleting the second row and column, as expected. We easily check that the eigenvalues for this restriction are just the diagonal elements: 0 with multiplicity k-1, and $-v_0 \neq 0$ with multiplicity k.

Notice that the multiplicity $k-1 = \dim S_h$ of 0 has its minimum possible value, all non zero eigenvalues equal to $-v_0$ are real and their invariant subspace is seen to be complementary to $T_{x_0}S_h$. We say that S_h is a normally hyperbolic submanifold of $\bar{E}_h$, and under these conditions a theorem by Fenichel |2| states that there exists a neighborhood of S_h^+ which is positively invariantly, fibered by a family of stable submanifolds (in correspondence with each equilibrium point on S_h^+). By symmetry, a similar statement is valid for S_h^- with negatively invariant submanifolds.

It is clear that the dimension of the invariant submanifolds associated with each equilibrium point, must equal the codimension of S_h in $\bar{E}_h$. We can check that this is exactly k, as asserted.

Q.E.D.

We remark that Malkin |5| already knew a weak form of the result by Fenichel we referred to above. Aulbach has considered generalizations of the same result (see |1| and references therein).

We may interpret Theorem 1 by saying that given any direction of asymptotic approach to infinity, there exists a k-dimensional submanifold of $\bar{E}_h$, consisting of orbits which escape to infinity in the given direction.

We end this section with a remark about $r(\tau)$, where we recall that $r = \rho^{-1}$. We can easily check that $r'' = r(r\dot{r})^{\bullet} = r\ddot{I}/2$, so that the Lagrange-Jacobi identity gives

$$r'' = rU(q) + hr = U(Q) + 2hr$$

Since U is bounded away from zero on S, this means that $r'' > 0$. In other words, $r(\tau)$ is strictly convex as it was the case for $I(t)$. Hence, $r(\tau) \to \infty$ for $\tau \to +\infty$ if solution is defined for all τ. In this time scale, total collision $r \to 0$ still appers as a singularity. In fact, τ is a faster time variable, since it is of the order of $\ln|t|$ as $t \to \pm\infty$.

4. DESCRIPTION OF THE FLOW AT INFINITY, EXAMPLES. It turns out that the fictitious flow on the infinity manifold for $h > 0$ can be explicitly integrated. In this section, we give a geometrical description of said flow, ending with a description of the neighboring flow in $\bar{E}_h$ for two illustrative examples: the Kepler problem in 1 dimension, and the plane isosceles 3-body problem.

Our new result describing the flow on N_h for $h > 0$, can be summarized as follows.

PROPOSITION 2.- The fictitious vector field on N_h is almost gradient with respect to v. Except for singularities on N_h corresponding to lower order collisions, the complement $N_h \setminus S_h$ is a global stable manifold for the flow on N_h with respect to the attractor S_h^+, and a global unstable manifold for the repeller S_h^-. Given a point $Q_{+\infty} \in S$, the k-2 parameter family of solutions attracted by $Q_{+\infty}$ as $\tau \to \infty$ is described as

$$Q(\tau) = Q_{+\infty} \tanh a\tau + Q_0 \operatorname{sech} a\tau$$
$$v(\tau) = a \tanh a\tau$$

where $a = \sqrt{2h}$ and $Q_{+\infty} A Q_0^t = 0$.

<u>Proof</u>: If we let $\rho = 0$, then (5) shows that $v' = Q'AQ'^t \geq 0$. Looking at the equation for Q'', we see that $v'=0$ only at equilibrium points in S_h. This shows the flow is almost gradient

with respect to v. Since v increases, non equilibrium orbits on N_h go away from S_h^- and tend to S_h^+, unless they are interrupted by a non total collision singularity.

Combining $v' = Q'AQ'^t$ with the energy relation (6), we obtain

$$(7) \qquad v' = 2h - v^2 \geq 0$$

Let us now look at the equivalent system of differential equations (3). We see that if $\rho = 0$, the momenta are constant, say $p = p_0 \in \mathbb{R}^k$. Substitution in the equation for Q'gives

$$(8) \qquad Q' = -vQ + p_0A^{-1}$$

Integration of (7) with initial condition $v(0) = 0$ gives $v(\tau) = a \tanh a\tau$, as required. Substitution of $v(\tau)$ into (8) gives a first order linear equation which we solve as

$$Q(\tau) = p_0A^{-1}\tanh a\tau/a + Q_0 \operatorname{sech} a\tau$$

where we check that $Q_{+\infty} = a^{-1}p_0A^{-1}$. Recalling that $v = pQ^t$ and since $v(0) = 0$, we get $p_0Q_0^t = 0$. Substitution of $Q_{+\infty}$ gives

$$(9) \qquad Q_{+\infty} AQ_0{}^t = 0,$$

as asserted. To check that this is a k-2 parameter family of solutions, we see that given any $Q_{+\infty} \in S$ (9) restricts the initial condition Q_0 to be orthogonal to $Q_{+\infty}$ in the A-norm. That means it may be any point in the "equator" of S with respect to $Q_{+\infty}$, which has dimension k-2. If this equator intersects singularities of U in S, $Q_{+\infty}$ can not be reached from some Q_0, which means that orbits are interrupted.

When we let $\tau \to -\infty$, we check $Q_{-\infty} = -Q_{+\infty}$

Q.E.D.

We observe from system (5) that if $\rho = 0$, all the terms containing the potential U disappear. This means that in principle the flow on N_h does not have singularities. This is however an illusion, since Q is constrained for non fictitious states to the set S which avoids the singularities of the potential. This explains the interruption of orbits we took into account above. Of course, those corresponding to binary collisions can be continued with another one through regularization, as we will see in Example 2, below.

We consider now the illustrative examples

Example 1) Consider the Kepler problem in one dimension. The phase space is given by positions $x > 0$ and velocities or momenta $p \in \mathbb{R}$. System (1) and energy relation (2) are written as (Fig.1)

$$\dot{x} = p$$
$$y = -x^{-2}$$
$$p^2/2 = x^{-1}+h$$

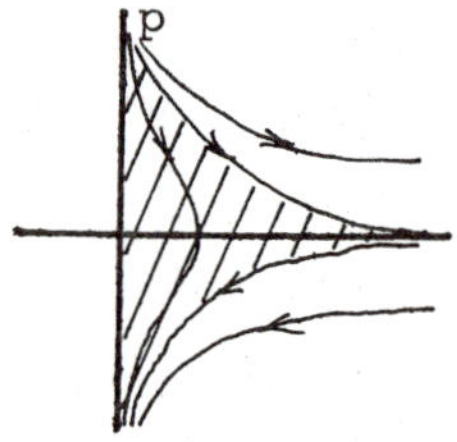

Fig. 1

The blow up at ∞ for $h > 0$ in this case amounts to $z=x^{-1}$ and $dt = z^{-1}d\tau$. The new energy relation becomes $p^2/2= z + h$, whose level lines are parabolas, as shown in Fig. 2.

Incidentally, we mention that the time scale used in this case, is the same we need when a regularization of binary collisions is performed. We would need different phase space variables for that. However, only for Kepler problems the total collapse and binary collision do coincide.

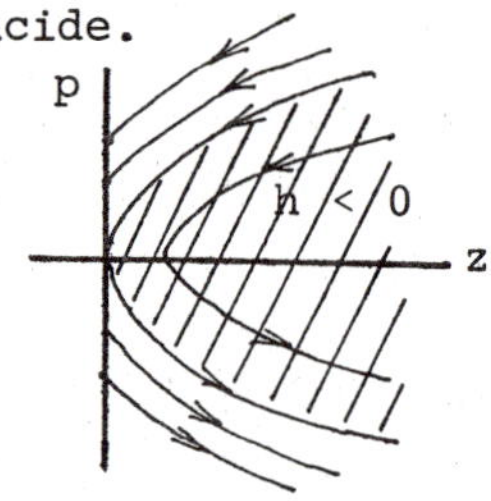

Fig. 2

Example 2) Consider now the (planar) isosceles three body problem. We are given a particle of mass α moving along an axis and two particles of unit mass symmetrically situated with respect to the axis. Using as Jacobi-like coordinates the semi-base $x > 0$ and the height $y \in \mathbb{R}$, the energy relation is

$$1/2\ pA^{-1}p^t = (2x)^{-1} + 2\ \ (x^2+ y^2)^{-1/2} + h, \text{ where}$$

$A = \operatorname{diag}(2, 2\alpha/(2+\alpha))$.

With regularization of binary collisions, the topology of N_h for $h > 0$ is a sphere S^2 where 4 points have been deleted. The total collision manifold $C\ |4|$ has exactly the same topology. Indeed the whole energy level with its added boundaries $E_h \cup N_h \cup C$ is topologically (S^2-4 points) $\times$ $[0,1]$. The flow on C is known to have 6 equilibrium points, while that on N_h has 2 curves $S_h^{\pm}$ of equilibria, as shown in Fig. 3.

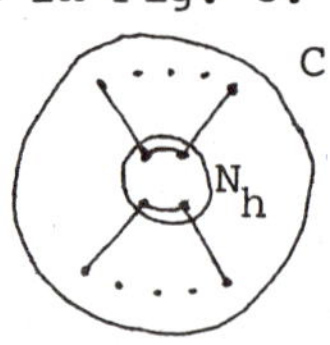

Fig. 3

Any asymptotic direction of escape in the half plane $x > 0$, $y \in \mathbb{R}$ is possible. The flow on N_h and nearby (with regularization of binary Collisions of course), is shown in Fig. 4, below. A few invariant manifolds on $\bar{E}_h$ are shown.

Fig. 4

5. FINAL REMARKS.- In Example 1) above, we see that trivially all the positive energy orbits for the Kepler problem are attracted (or repelled) to infinity. However, the situation can be very complicated for problems with more than 2 bodies as in Example 2). An interesting task that we will pursue in another paper, is the estimation of the basins of attraction or repulsion of S_h^+ and S_h^-.

Another interesting problem is the study of orbits escaping to ∞ in an oscillatory way, so that they do not approach any point in S_h.

BIBLIOGRAPHY

1. B. Aulbach, Approach to hyperbolic manifolds of stationary solutions, Lecture Notes in Mathematics 1017, Springer 1983, 56-66.

2. N. Fenichel, Geometric singular perturbation theory for ordinary differential equations, J. Differential Equations 31(1979) 53-98.

3. E. Lacomba, Variétés de l'infini pour une energie non nulle in mécanique céleste, C.R.A.S. Paris I295(1982) 503-506.

4. E. Lacomba and C. Simó, Boundary manifolds for energy surfaces in celestial mechanics, Celestial Mechanics 28(1982)37-48

5. I.G. Malkin, Theory of Stability of Motion (Russian), Moscow 1952.

6. R.Mc Gehee, Triple collision in the collinear three-body problem, Inventiones Math. 27(1974) 191-227.

7. H. Pollard, Celestial Mechanics, Carus Math. Monogr. #18, Mathem. Assoc. Amer. 1976.

8. C. Simó, Analysis of triple collision in the isosceles problem, in Classical Mechanics and Dynamical Systems, Marcel Dekker, New York 1981, 203-224.

DEPARTMENT OF MATHEMATICS
UNIVERSIDAD AUTONOMA METROPOLITANA, IZTAPALAPA
P.O. BOX 55-534, 09340 MEXICO, D. F.

Contemporary Mathematics
Volume 58, Part III, 1987

HOMOCLINIC ORBITS, HYPERBOLIC DYNAMICS AND DIMENSION OF CANTOR SETS

J. Palis

ABSTRACT. We discuss results about one-parameter families of diffeomorphisms: the creation of a homoclinic orbit of tangency and how frequently the transformation is hyperbolic when this tangency unfolds to determine transverse homoclinic orbits. The results depend on the dimension of certain Cantor sets. We pose several questions concerning this phenomenon.

About 1899, when studying the restricted three-body problem, Poincaré detected the existence of homoclinic orbits and he expressed much amazement about the dynamical complexity that such orbits seemed to imply [6]. To recall this concept, let f be a diffeomorphism of a manifold M and p a fixed point for it. We say that p has a homoclinic orbit through x if $\lim f^n(x) = p$ for both $n \to \infty$ and $n \to -\infty$. A similar definition holds when p is periodic of period m by just considering f^m. We are particularly interested in the case where p is hyperbolic: $df(p)$ has no eigenvalue with norm one. In this situation, the set of points $y \in M$ such that $f^n(y) \to p$ for $n \to \infty$ $(n \to -\infty)$ is an immersed submanifold called stable (unstable) manifold of p. Homoclinic orbits associated to p are then the intersection (transversal or not) of its stable $W^s(p)$ and unstable $W^u(p)$ manifolds. (Similarly for periodic points).

© 1987 American Mathematical Society
0271-4132/87 $1.00 + $.25 per page

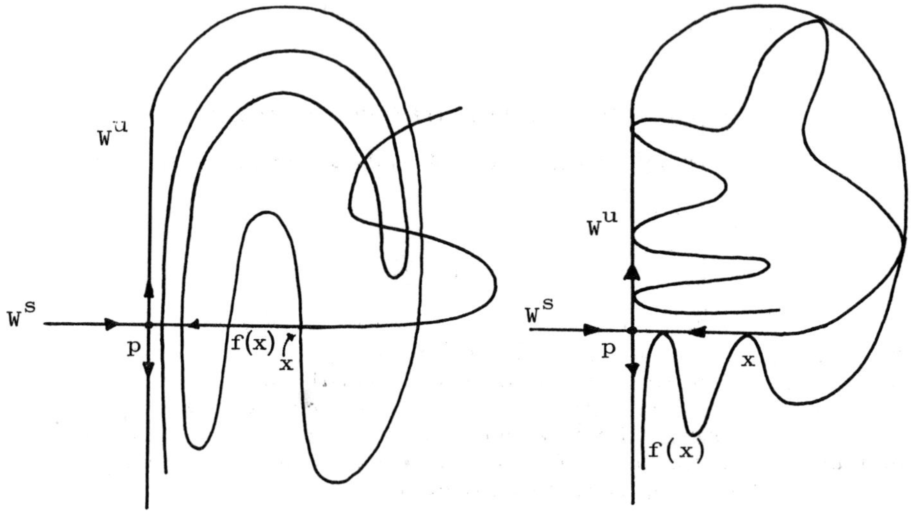

These figures give an idea of how complicate the dynamics of f (space of orbits) looks like near a homoclinic orbit. By 1935, Birkhoff showed that every transversal homoclinic orbit is accumulated by hyperbolic periodic points [1]. Actually, as proved by Smale some 30 years later, every transversal homoclinic orbit is part of an f-invariant compact hyperbolic set Λ in which the set of periodic orbits is dense (a "horseshoe"). That is, $f(\Lambda) = \Lambda$ and the tangent bundle $T_\Lambda M$ can be written $T_\Lambda M = E^s \oplus E^u$, where E^s and E^u are invariant by df, and $\|df \mid E^s\| < \lambda < 1$ $\|df^{-1} \mid E^u\| < \lambda < 1$ for some Riemannian metric on M. For a "linear" horseshoe as in the figure

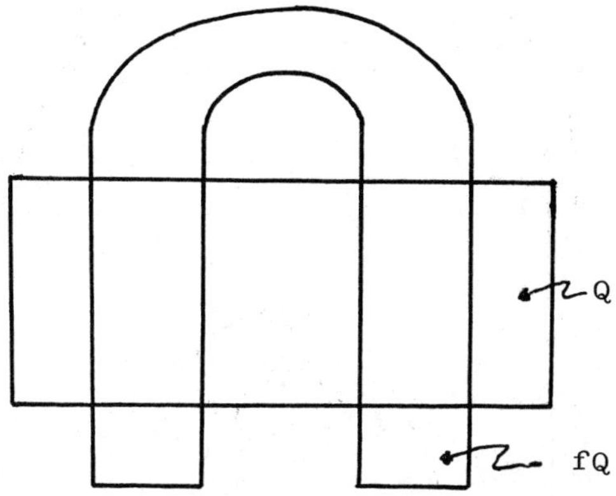

the maximal invariant set $\Lambda = \bigcap_{n\in\mathbb{Z}} f^n Q$ is a hyperbolic Cantor set (product of an "horizontal" and a "vertical" Cantor set). Through the points of Λ there exists an f-invariant foliation with leaves tangent to E^s: it is called stable foliation and denoted $\mathfrak{F}^s(\Lambda)$; dually, one has the unstable foliation $\mathfrak{F}^u(\Lambda)$. In the linear case, these foliations are made of horizontal and vertical lines, respectively. It is already remarkable that such a (hyperbolic) horseshoe is persistent in the following sense: if g is close to f in the C^1 topology then there is a (hyperbolic) horseshoe Λ_g for g. Moreover, $f|\Lambda$ and $g|\Lambda_g$ have the same dynamics: there is a homeomorphism $h: \Lambda \to \Lambda_g$ such that $hf(x) = gh(x)$ for $x \in \Lambda$. We can then ask: is it possible to detect a horseshoe in the presence of a transversal homoclinic orbit? The following figure gives a clear indications of how this can be done by considering the rectangle Q and a large iterate of it $f^n Q$:

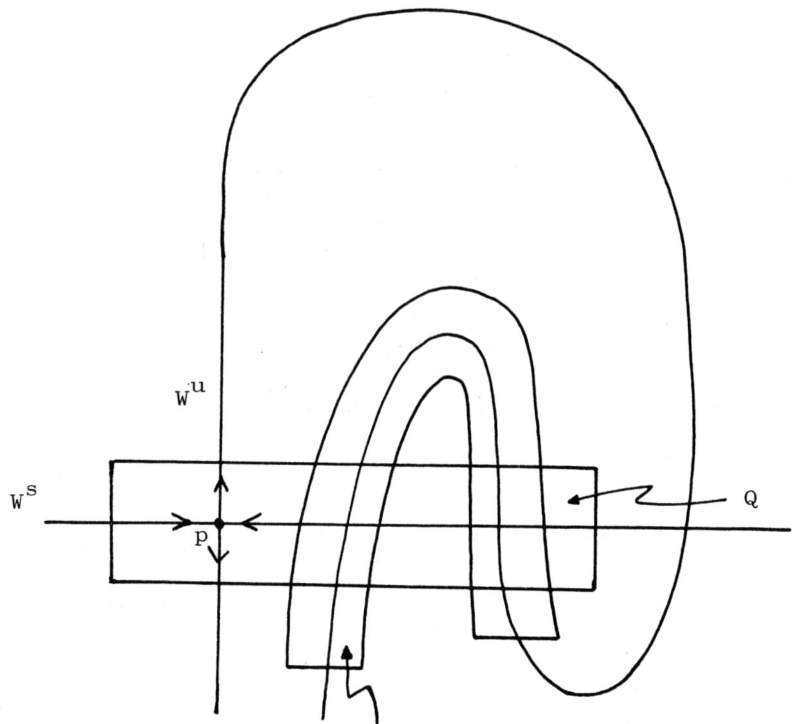

Imagine now for a family of transformation f_μ, $\mu \in R$, that we have no homoclinic orbits for $\mu < 0$ (and so no horseshoes) and we create a homoclinic tangency at $\mu = 0$. If this orbit of tangency becomes transversal for $\mu > 0$ (we say it unfolds generically) then we are "adding" at least a horseshoe to the dynamics of f_μ, $\mu \leq 0$. This horseshoe has in its own infinitely many hyperbolic periodic points as well as stable and unstable foliations. Thus by unfolding the initial homoclinic tangency as to create transversal homoclinic orbits, we obtain for infinitely many values of the parameter μ near zero new orbits of tangency between stable and unstable leaves. (The leaves may or may not be associated to periodic orbits).

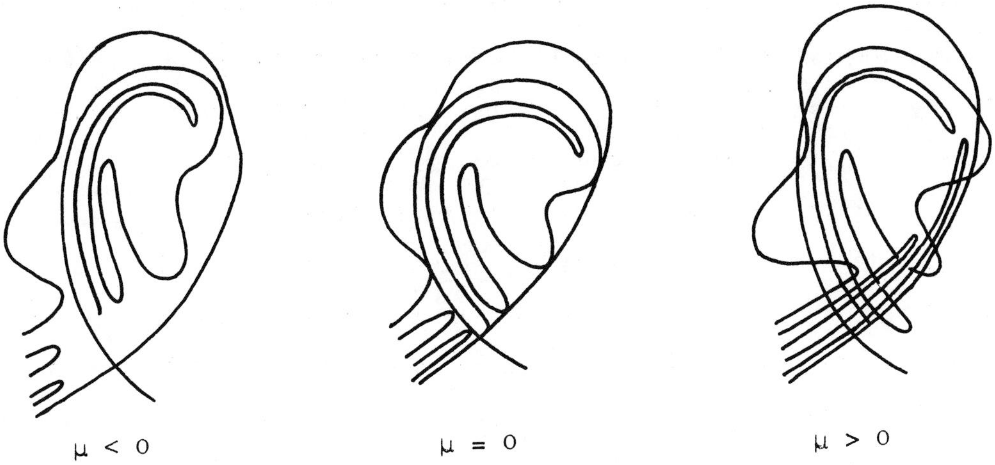

For these values of the parameter the dynamics of f_μ changes considerably with small variations of μ, as in the case of the initial homoclinic orbit at $\mu = 0$. The main reason is that again due to the tangency, f_μ is not hyperbolic, which means that the limit set L_μ of f_μ is not hyperbolic. Recall that $L_\mu = L^- \cup L^+$ where L^- is the closure of the union of the α-limit sets of the f_μ-orbits and L^+ of the ω-limit sets (y is in the α-limit set of the f_μ-orbit of x if there is a sequence $n_i \to \infty$ such that $f_\mu^{-n_i}(x) \to y$). Moreover, due to a remarkable result of Newhouse [3], there are intervals I_n arbitrarily close to zero in the μ-axis such that for a dense subset of μ's in I_n, f_μ has infinitely many sinks or sources (periodic attractors or repellors). Again, this is an unstable phenomenon. For these reasons, the analysis of the (global) dynamical structure of a transformation when we unfold a homoclinic tangency may seem rather intractable. However, for a large class of one-parameter families of diffeomorphisms going through a homoclinic tangency, the transformation is persistently hyperbolic in large portions of the parameter space (in the sense of Lebesgue measure) near the value corresponding to the initial homoclinic tangency. Thus, one mostly sees hyperbolicity (a "large" horseshoe containing the fixed point) and, consequently much stability of the dynamics.

Our purpose here is to discuss recent results in this direction, obtained with F. Takens [4,5], and pose several related problems. The results draw a correlation between fractional dimensions (Hausdorff dimension, limit capacity) of stable and unstable foliations defined on Cantor sets (horseshoes) and hyperbolic dynamics.

§1. BASIC CONCEPTS. Let M be a closed (compact without boundary) smooth manifold. Recall that if a diffeomorphism $f: M \circlearrowleft$ is hyperbolic (i.e., the limit set $L(f)$ is hyperbolic) then $L(f) = L_1 \cup \dots \cup L_k$, where each L_i, $1 \leq i \leq k$, is f-invariant, closed and transitive (has a dense orbit); it follows that the periodic points are dense in L_i. Each L_i is called a basic set. Through each point x of L_i we can construct the stable and unstable manifolds $W^s(x)$, $W^u(x)$ (immersed $\mathbb{R}^s$, $\mathbb{R}^u$) defined as follows: $y \in W^s(x)$ if $d(f^n(x), f^n(y)) \to 0$ as $n \to \infty$ and $z \in W^u(x)$ if $d(f^n(x), f^n(z)) \to 0$ as $n \to -\infty$. Put together, $W^s(x)$ for $x \in L_i$ form the stable foliation $\mathfrak{F}^s$ of L_i. Similarly, we have the unstable foliation $\mathfrak{F}^u(L_i)$ whose leaves are $W^u(x)$ for $x \in L_i$. The basic set L_i is called an attractor or a repellor if the leaves of $\mathfrak{F}^s(L_i)$ or of $\mathfrak{F}^u(L_i)$ cover a neighborhood of L_i; otherwise, L_i is of saddle-type. Notice that if f is of class C^2 and $\dim M = 2$, then $\mathfrak{F}^s(L_i)$ and $\mathfrak{F}^u(L_i)$ can be extended to f-invariant $C^{1+\varepsilon}$ foliations defined in a full neighborhood V_i of L_i; i.e. the projection along leaves into a (local) transversal section is C^1 with Hölder continuous derivative (actually, the map that associates to each point $x \in V_i$ the tangent space to the leaf through x is $C^{1+\varepsilon}$). Another relevant concept here concerning basic sets, is that of a <u>cycle</u>: there is a cycle for $L_{i_1}, \ldots, L_{i_k}$ if there are pair of points $x_j, y_j \in L_{i_j}$ such that $W^u(x_j) \cap W^s(y_{j+1}) \neq \phi$ for $1 \leq j < k$ and $W^u(x_k) \cap W^s(y_1) \neq \phi$.

On the other hand, for a (compact) subset A of a metric space X the following notions of "dimensions" can be introduced. For each ε, let $N(\varepsilon)$ be the minimal number of ε-balls needed to cover A. Then the limit capacity $d(A)$ is defined as

$$d(A) = \limsup_{\varepsilon \to 0} \frac{\operatorname{Log} N(\varepsilon)}{-\operatorname{Log} \varepsilon}$$

(the concept seems to be due to Kolmogorov). Let us now define the Hausdorff dimension of A: for each real number $p > 0$, we consider

$$H(p) = \lim_{\varepsilon \to 0} \inf_{\mathcal{G}} \Big(\sum_{V_i \in \mathcal{G}} \rho(V_i)^p \Big),$$

where for each $\varepsilon > 0$ the inf is taken over all possible countable covers of A whose elements V_i have diameter $\rho(V_i)$ less than ε. The Hausdorff dimension $HD(A)$ is then defined as $\inf\{p;\ H(p) = 0\}$. In general $HD(A) \leq d(A)$, but we shall stress in the sequel that in our context we will deal with sets A for which $HD(A) = d(A)$. If $HD(A) = d(A)$ and $HD(B) = d(B)$ the $HD(A \times B) = H(A) + H(B) = d(A) + d(B) = d(A \times B)$. Notice for example that for $0 < \alpha < 1$ the α-middle third Cantor set K_α in $[0,1]$, where α measures the "gap" in the middle, we have

$$HD(K_\alpha) = d(K_\alpha) = \frac{\operatorname{Log} 2}{\operatorname{Log} 2 - \operatorname{Log}(1-\alpha)}.$$

Now, let f be a hyperbolic diffeomorphism of class C^2 of a closed surface M^2 (i.e., the limit set $L(f)$ is hyperbolic) and let L_i be one of its basic set. We assume that L_i is of saddle-type (neither an attractor nor a repellor) and that it is not made of a unique periodic orbit. For any $x \in L_i$ and any $y \in W^s(x)$ take a small segment Σ_y transversal to $W^s(x)$ through y and consider the point set intersection of the stable leaves of L_i with Σ which we call A_Σ. (Actually we take $\Sigma \cap W^s_\delta(z)$ for $z \in L_i$ and δ small, where $W^s_\delta(z)$ is the (intrinsic) δ-disk centered at z). The <u>stable limit capacity</u> of L_i is then defined as the limit capacity $d^s(L_i)$ of A_Σ. Of course, one has to show that $d(A_{\Sigma'}) = d(A_\Sigma)$ for any other cross section Σ' as above. This follows from the fact that $\mathfrak{F}^s(L_i)$ is $C^{1+\varepsilon}$ and so it defines diffeomorphisms between sections (by going along leaves) and the above <u>dimensions</u> are <u>invariant</u> by <u>Lipschitz homeomorphisms</u>. Notice that A_Σ is a Cantor set which we will call <u>regular</u> (because it is "generated" by $C^{1+\varepsilon}$ expanding maps, i.e. maps with derivative bigger than 1 everywhere). See Section 3. Similarly we can define the <u>unstable limit capacity</u> $d^u(L_i)$.

§2. RESULTS. As indicate before, we consider the set $\mathcal{G}$ of parametrized families $\{f_\mu ; \mu \in \mathbb{R}\}$ of diffeomorphisms of a closed surface such that

i) f_μ is hyperbolic for $\mu < 0$, and as $\mu \nearrow 0$ its limit set $L(f_\mu)$ converges (in the Hausdorff metric) to a hyperbolic set L; in particular, $L_{\mu,i} \nearrow L_i$ where the $L_{\mu,i}$ are the basic sets of f_μ and $L = L_1 \cup \ldots \cup L_k$,

ii) the limit set of f_0 consists of $L = \lim_{\mu \nearrow 0} L(f_\mu)$ together with a homoclinic tangency which is quadratic and unfolds generically; except for that there are no other cycles.

These conditions may seem rather restrictive, but due to previous work of Newhouse and the author (see [8]) we are led to believe that they correspond to a typical way of changing or "bifurcating" the dynamics of f_μ if we assume that "initially" f_μ is hyperbolic (i.e. $L(f_\mu)$ is hyperbolic say for $\mu < 0$). The following figures describe some of these cases.

(a)

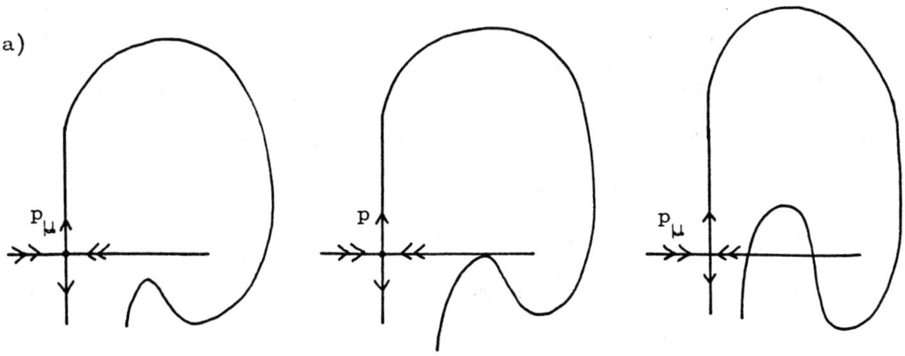

(b)

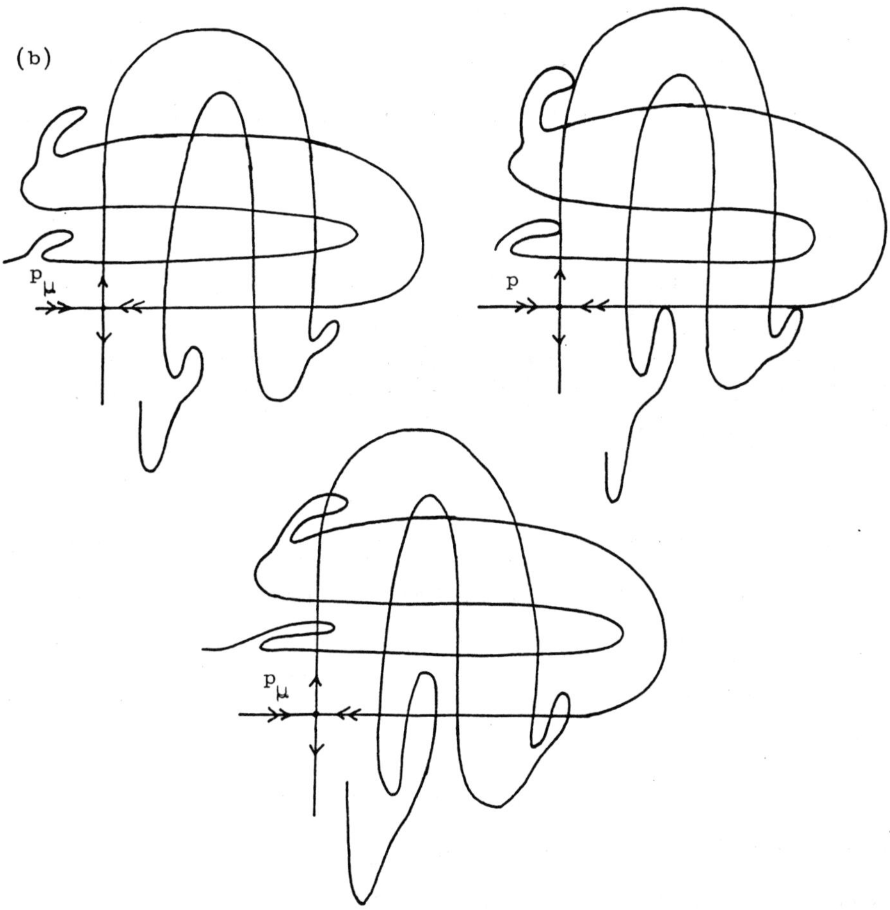

In (a) the homoclinic tangency is associated to a basic set that consists of a fixed point p. The double arrow in the stable manifold against the single arrow in the unstable means that the product of eigenvalues of $df(p)$ is less than one. In such a case each component of $W^s(p)-\{p\}$ is called "strong" and of $W^u(p)-\{p\}$ "weak". Actually instead of the product of the eigenvalues what really counts here is that a strong component accumulates in itself but not a weak one.

If a weak component accumulates in itself, like in (a') below, then one shows that we can not achieve the homoclinic tangency say at $\mu = 0$ without first creating horseshoes: for small values of $\mu < 0$ we already have horseshoes and thus we must have infinitely many homoclinic tangencies ("cascade of bifurcations") for $\mu < 0$!

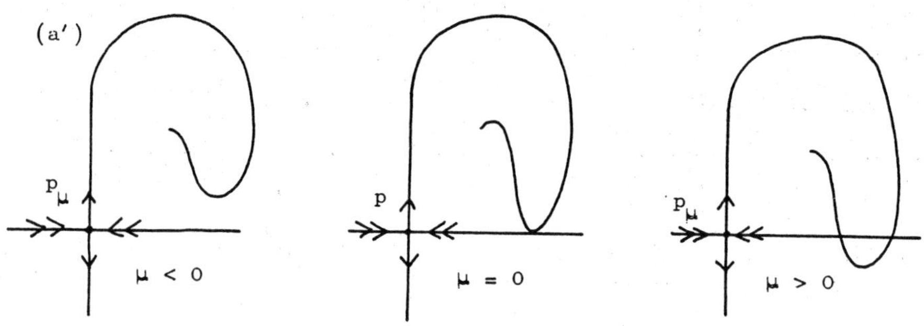

A very well known question is whether one has "strange" attractors (in particular nonperiodic attractors) for values of μ near zero; see Question 6 in the next section. We stress that the results we will state below do not address such a case (a') as indicated in the figure.

For $\{f_\mu\}$ in $\mathcal{Q}$ as above, let d^s and d^u denote the stable and unstable limit capacities of the basic set to which the homoclinic tangency at $\mu = 0$ is associated. Let also $B_{(o,\delta)}$ denote the set of values $\mu \in (o,\delta)$ for which f_μ is not hyperbolic ($L(f_\mu)$ not a hyperbolic set) and $m(B_{(o,\delta)})$ its Lebesgue measure.

<u>Theorem</u> [5]. If $d^s+d^u < 1$, then

$$\lim_{\delta\to 0} \frac{m(B_{(o,\delta)})}{\delta} = 0.$$

A similar result can be stated for families $\{f_\mu\}$ such that f_μ is hyperbolic for $\mu < 0$ as before but instead of a homoclinic tangency at $\mu = 0$ we create a cycle involving more than one basic set as in the following figure:

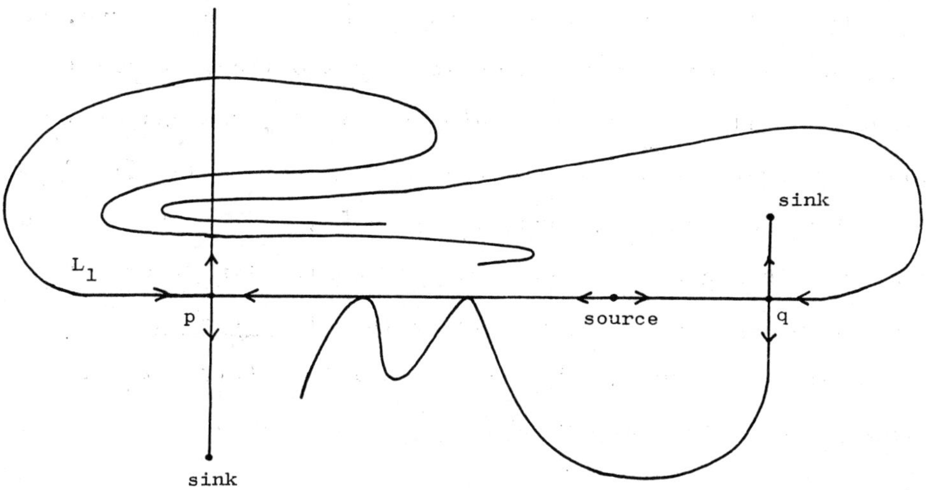

In this figure we have two basic sets involved in the cycle: a horseshoe L_1 and a fixed point q; the other basic sets are two sinks (point attractors) and a source (point repellor). In such cases we define d_c^s to be maximum of the stable capacities of the basic sets involved in the cycle; dually for d_c^u.

<u>Theorem</u> [4]. If $d_c^s+d_c^u < 1$, then

$$\lim_{\delta\to 0} \frac{m(B_{(o,\delta)})}{\delta} = 0.$$

Newhouse and the author first obtained results of this nature in [8] for the particular case where each basic set consists of a periodic orbit. (Also, in that work we were only able to show that $\liminf m(B_{[0,\delta]})/\delta = 0$ instead of the full limit).

§3. QUESTIONS. The first question we want to pose is a kind of converse to the previous results.

1. <u>Conjecture</u>. For generic families $\{f_\mu\} \in \mathcal{C}$ as considered in Section 2, if $d^s + d^u > 1$ then $\liminf m(B_{[0,\delta]})/\delta > 0$. Similarly for the case of cycles if $d_c^s + d_c^u > 1$.

Motivated by the previous results and this conjecture, we were led to questions about Cantor sets on the line. Actually, we are interested in "regular" Cantor sets: they are generated in standard way from say ℓ alternate subintervals I_i of $[0,1]$ and $C^{1+\epsilon}$ surjective expanding maps $J_i : I_i \to [0,1]$. If the J_i are affine maps, the resulting Cantor set is called linear or affine. For regular Cantor sets C_1, C_2 we define its <u>difference</u> as $C_1 - C_2 : \{p-q;\ p \in C_1 \text{ and } q \in C_2\} = \{\mu \in \mathbb{R}\ ;\ C_1 \cap (\mu + C_2) \neq \emptyset\}$. Here $\mu + C_1$ indicates the translation of C_1 by μ.

2. <u>Question</u>. Is it true that for linear Cantor sets we have either $m(C_1 - C_2) = 0$ or else $C_1 - C_2$ contains an interval?

3. <u>Question</u>. Similarly for regular Cantor sets.

Notice that Question 2 is true for $C_\alpha - C_\alpha$ where C_α is the standard α-middle-third Cantor set, 

. We have $C_\alpha - C_\alpha = [-1,1]$ for $\alpha \le 1/3$ and $m(C_\alpha - C_\alpha) = 0$ for $\alpha > 1/3$. We do not know the answer in general for $C_\alpha - C_\beta$. Observe that if $d(C_1) + d(C_2) < 1$ then $m(C_1 - C_2) = 0$ (see [4]). On the other hand for almost all $\rho \in \mathbb{R}$ $m(C_1 - \rho C_2) > 0$ if $d(C_1) + d(C_2) > 1$ (Marstrand [2]). Here we are using that $HD(C) = d(C)$ for a regular Cantor set C as pointed out by L.S. Young and F. Ledrappier (F. Takens observed that this is true even if the generating maps are just C^1 instead of $C^{1+\varepsilon}$). Notice that $m(C_\alpha - C_\alpha) = 0$ for $\frac{1}{3} < \alpha < \frac{1}{2}$ and yet $2d(C_\alpha) > 1$ ($\rho=1$ is an exceptional value in this case).

4. Question. Can we define limit capacity and Hausdorff dimension for the stable (unstable) foliation of a basic set in higher dimensions? Notice that these foliations are not (transversally) differentiable in general. See Section 1.

5. Question. To what extent can the previous results be carried out in higher dimensions? F. Takens and the author made some initial progress in this direction. How about Conjeture 1? One can also ask questions for families $\{f_\mu\} \in \mathcal{G}$ about the entropy and stable and unstable dimensions for f_μ, $\mu > 0$ and small. We had some discussions with L.S. Young on this topic. Roughly speaking, they tend to increase with μ for small $\mu > 0$.

6. Question (Hénon, Newhouse). What can be said about the dynamics of f_μ for small $\mu > 0$ if for $\mu < 0$ one also has a cascade of bifurcations? See Section 2, case (a'). Besides Newhouse's intervals in the parameter space (for which there are dense subsets such that the corresponding transformations exhibit infinitely many sinks or sources), so far there seems to be little progress in this question. This kind of bifurcation also appeared in the work Gravilov and Silnikov as well as that of Newhouse and the author. Hénon suggested the existence of a "strange attractor" for a family of quadratic maps. Later Newhouse asked about "strange attractors" for the more general families described above and values of μ near zero. A relevant problem here is whether there are sets of positive measure near $\mu = 0$ for which f_μ has no periodic attractors with orbits passing near the homoclinic tangency.

REFERENCES

[1] G.D. Birkhoff, Nouvelles recherches sur les systèmes dynamiques, Mem. Pont. Acad. Sci. Novi Lyncaei, 1 (1935), 85-216.

[2] J.M. Marstand, Some fundamental geometrical properties of plane sets of fractional dimensions, Proc. London Math. Soc. (3) 4 (1954), 257-302.

[3] S. Newhouse, The abundance of wild hyperbolic sets and non-smooth stable sets for diffeomorphisms, Publ. Math. IHES 50 (1979), 101-151.

[4] J. Palis & F. Takens, Cycles and measure of bifurcation sets for two-dimensional diffeomorphisms, Invent. Math. 82, 1985.

[5] J. Palis & F. Takens, Hyperbolicity and creation of homoclinic orbits, IMPA, preprint, 1985.

[6] H. Poincaré, Les méthodes nouvelles de la Mécanique Celeste III, Gauthiers-Villars, 1899.

[7] S. Smale, Diffeomorphisms with many periodic points, Differential and Combinatorial Topology, Princeton University Press, 1965.

[8] S. Newhouse & J. Palis, Cycles and bifurcation theory, Astérisque, 31 (1976), 44-140.

INSTITUTO DE MATEMÁTICA PURA E APLICADA
Estrada Dona Castorina 110,
CEP 20460 - Rio de Janeiro, RJ - Brazil

Contemporary Mathematics
Volume 58, Part III, 1987

ON THE ESTIMATE OF NUMBER OF SOLUTIONS FOR THE NONLINEAR BOUNDARY VALUES PROBLEM

Vladimir Schuchman

INTRODUCTION. In this paper we study elliptic boundary value problems of the form

$$\begin{aligned} Au &= \lambda f(x,u) \quad \text{in } \Omega \\ u &= 0 \quad \text{on } \partial\Omega \end{aligned} \tag{1}$$

where Ω is a bounded domain in R^n of class $C^{2+\mu}$ for some $\mu \in (0,1)$. Moreover

$$Au = -\sum_{jk=1}^{n} a_{jk} \mathcal{D}_j \mathcal{D}_k u + \sum_{j=1}^{n} a_j \mathcal{D}_j u + a_0 u$$

with $a_{jk} = a_{kj}$, a_{jk}, a_j, $a_0 \in C^1(\overline{\Omega}, R)$. We assume that

$$a_0 \geq 0$$

and that there exists a constant $\underline{C}$ such that

$$\sum_{j,k=1}^{n} a_{jk} \zeta_j \zeta_k \geq \underline{C} |\zeta|^2 \quad \forall \zeta \in R^n \tag{2}$$

As for nonlinearity we assume that

$$f \in C^{\mu,1}(\overline{\Omega} \times R^+, R^+)$$

that is, f is μ-Holder continuous in $x \in \overline{\Omega}$ and continuously differentiable in $\zeta \in R^+$. Moreover, we suppose that

(i) $f(x,0) > 0 \quad \forall x \in \overline{\Omega}$

(ii) There exist constants K and $\alpha > 0$ such that

$$0 < \mathcal{D}_2 f(x,\zeta) \leq K/(1+\zeta)^{1+\alpha}$$

for all $x \in \overline{\Omega}$ and $\zeta \in R^+$.

Under those assumptions it is easy to see that problem (1) has a unique solution of class $C^{2+\mu}$ for sufficiently small λ. Under the same assumptions in [6] was proved that there exists a unique solution for sufficiently large λ.

Problems of the form (1) arise in many applications, e.g. in chemical

© 1987 American Mathematical Society
0271-4132/87 $1.00 + $.25 per page

engineering questions or in the theory of nonlinear heat propagation, and they attracted a considerable amount of interest during the last two decades.

It follows obviously from the maximum principle that every solution of (1) with $\lambda > 0$ is positive. Moreover it is a well known consequence of the method of sub-and-super solutions that there exists a $\lambda^* \varepsilon (0,\infty]$ such that (1) has for each $\lambda \varepsilon (0,\lambda^*)$ minimal solution and no solutions for $\lambda > \lambda^*$. In general, however, problem (1) will have more than one solution for $\lambda \varepsilon (0,\lambda^*)$ (cf[1,4]). Theorem from [6] shows that -given the above hypotheses- first of all $\lambda^* = \infty$ and secondly, problem (1) has a unique solution for large values of λ.

In this paper we give some estimate for the number of distinct solutions for (1), i.e. there exists function $N(\lambda)$ which depends on $f(x,u)$ such that problem (1) has at least $N(\lambda)$ distinct solutions for every $\lambda > 0$, (theorem 2). Also we give some simple result which is coincided with standard results about maximum of solutions of (1), namely there exists compact $K \subset \Omega$ such that maximum of every solution of (1) for every $\lambda > 0$ is achieved in K (theorem 1). Last section contains some remarks about Semenov's approximation. There we proved also the existence of function $f(u)$ such that problem (1) has at least N distinct positive solutions for $\lambda = \lambda_0$ (λ_0 is fixed positive) (theorem 3).

PROOF OF THE THEOREMS 1 AND 2. From assumptions (i)-(ii) it obviously follows that

(iii) There exists constants C_1 and $C_2 > 0$ such that

$$0 < C_1 \leq f(x,\zeta) \leq C_2 \quad \text{for } x \varepsilon \bar{\Omega}, \zeta \varepsilon R^+ .$$

Let e be the solution of the following problem

$$Ae = 1 \text{ in } \Omega \quad e = 0 \text{ on } \partial\Omega$$

Then $e > 0$ in Ω and exists $x_0 \varepsilon \Omega$ such that

$$e(x_0) = \sup_{\Omega} e(x)$$

First of all from maximum principle follows that

$$\lambda C_1 e(x) \leq u_\lambda(x) \leq \lambda C_2 e(x) \tag{3}$$

where $u_\lambda(x)$ is a solution of (1)

Let $\mathcal{D}_\varepsilon$ be a neighborhood of $\partial\Omega$ such that $e(x) < \varepsilon$ in $\mathcal{D}_\varepsilon$. Then $u_\lambda(x) \leq \lambda C_2 \varepsilon$ in $\mathcal{D}_\varepsilon$ and $u_\lambda(x_0) \geq \lambda C_1 e(x_0)$. If $\varepsilon > 0$ is sufficiently small (such that $C_2 \varepsilon < C_1 e(x_0)$), then value of $u_\lambda(x_0)$ at the point x_0 will be bigger than $\sup_{\mathcal{D}_\varepsilon} u\,(x)$ for all $\lambda > 0$ and from here follows that

$$\sup_{\Omega} u_\lambda(x) = \sup_{\Omega-\mathcal{D}_\varepsilon} u_\lambda\,(x) \tag{4}$$

Thus we proved the following:

THEOREM 1. Under the above hypotheses i)-iii) there exists compact $K \subset \Omega$, $K \cap \partial\Omega = \phi$ such that every solution of (1) for every $\lambda > 0$ has its maximum in K.

Let's note that if there are two solutions for problem (1) with distinct values at the point x_0 then there are distinct.

Now the question is how many solutions has problem (1) for middle values of λ.

If $G(x,y)$ demotes the Green's function for problem (1), then

$$u_\lambda(x) = \lambda \int_\Omega G(x,y)\, f(y,u_\lambda(y))dy \tag{5}$$

or

$$u_\lambda(x_0) = \lambda \int_\Omega G(x_0,y)\, f(y,u_\lambda(y))dy =$$

$$= \lambda \int_\Omega G(x_0,y)\, f(x_0,u_\lambda(x_0))\, \frac{f(y,u_\lambda(y))}{f(x_0,u_\lambda(x_0))}\, dy \tag{6}$$

and last expresion we can estimate by using the following property of $f(x,\zeta)$:

(iv) There exists constants $\tilde{C}_1, \tilde{C}_2 > 0$ such that

$$0 < \tilde{C}_1 \le \frac{f(y,\eta)}{f(x_0,\zeta)} \le \tilde{C}_2 < \infty \quad \text{for all } y \in \bar{\Omega},\ \zeta,\eta \in R^+$$

where $\tilde{C}_1 = \frac{C_1}{C_2}$; $\tilde{C}_2 = \frac{C_2}{C_1}$.

From here we obtain

$$\tilde{C}_1\, c \le \int_\Omega G(x_0,y)\, \frac{f(y,u_\lambda(y))}{f(x_0,u_\lambda(x_0))} dy \le \tilde{C}_2\, c \tag{7}$$

where

$$c = \int_\Omega G(x_0,y)dy = e(x_0)$$

By using (6) we have

$$\lambda\tilde{C}_1 c\, f(x_0, u_\lambda(x_0)) \le u_\lambda(x_0) \le \lambda\tilde{C}_2 c\, f(x_0,u_\lambda(x_0)) \tag{8}$$

By using now $\rho = u_\lambda(x_0)$ and $f(\rho) = c\, f(x_0, \rho)$ we have

$$f_2(\rho) = \frac{\rho}{f(\rho)\tilde{C}_2} \leq \lambda \leq \frac{\rho}{f(\rho)\ \tilde{C}_1} = f_1(\rho) \qquad (9)$$

and $\lambda(\rho)$ will be always between $f_1(\rho)$ and $f_2(\rho)$.

Let's note then from (9) one can derive solvability of (1) for all $\lambda > 0$. It was proved independently in [6].

Now let N_1 and N_2 be the following sets

$$N_1 = \{(\rho,\zeta) \mid \zeta = f_1(\rho)\}$$

$$N_2 = \{(\rho,\zeta) \mid \zeta = f_2(\rho)\}$$

and N^i_λ for $i = 1,2$ be a set of disjoint intervals or points on the line $\zeta = \lambda$ such that $(\rho_1\ \rho_2) \in N^i_\lambda$ iff $f_i(\rho) \geq \lambda$ on the interval $\rho_1\ \rho_2$.

Now let A be the following set

$$N^1_\lambda \setminus (N^1_\lambda \cap N^2_\lambda)$$

which contains a number of disjoint intervales $(\rho_1'\ \rho_2')$ such that one of the two possibilities hold

i) There exists an interval $(\rho_1'\ \rho_2) \in N^1_\lambda$ and an interval (maybe a point) $(\rho_2'\ \rho_3) \subset (\rho_1', \rho_2)$ and $(\rho_2', \rho_3) \in N^2_\lambda$ or

ii) There exists an interval $(\rho_1, \rho_2') \in N^1_\lambda$ and an interval (or point)$(\rho_3, \rho_1') \subset (\rho_1, \rho_2')$ and $(\rho_3, \rho_1') \in N^2_\lambda$.

Let now $N(\lambda)$ be a number of elements in the set A .

From connectedness of the set $\{\lambda,\ u_\lambda(x_0)\}$ or $\{\lambda, \rho\}$ [5] we can see that the following theorem is valid.

THEOREM 2. Under the above assumptions-problem (1) has at least $N(\lambda)$ distinct solutions.

REMARKS. 1. It is easy to see that under our assumptions $N(\lambda)$ has only finite values and $N(\lambda) = 1$ for sufficiently large and sufficiently small λ . Also always $N(\lambda) \geq 1$.

2. To distinct solution of problem (1) as usual is used $\|u\|_\infty$ and not value of

u at the specific point. First of all if we have problem (1) in the ball, operator $A = -\Delta$ and f does not depend on x, then all solutions of problem (1) will have radial symetry [3] and $u(0) = \|u\|_\infty$. Secondly, as follows from Theorem 1 $u(x_0)$ can not be very far from $\|u\|_\infty$.

3. If on interval $(\rho_1\ \rho_2)$ there exists at least one common point for sets N_1 and $\{\zeta = \lambda\}$ and N_2 and $\{\zeta = \lambda\}$ then there exists at least one solution of (1) such that $\rho_1 < u(x_0) < \rho_2$.

4. By using Bony's maximum principle [2] instead of the classical Hopf maximum principle we can allow $\mu = 0$, provided we mean by a solution a function $u \in W_p^2(\Omega)$ for some $p > n$ which satisfy (1) almost everywhere.

5. In the case of $A = -\Delta$, Ω is a ball and f does not depend on x maximum of every solutions is achieved at the origin. In our case it is not valid but at least as shows theorem 1 maximum is achieved out of compact neighbourhood of $\partial\Omega$. We think this is the best possible result that could be achieved.

ABOUT SEMENOV'S APPROXIMATION. In some cases it is well known that for the estimate of the number of solutions for (1) it is sufficient to consider the following problem

$$v = \lambda f(v) \tag{10}$$

In such a cases the behavior of the set $\{\lambda\ u_\lambda\}$ of solutions for (1) resembles the behavior of the set $\{\lambda\ , v\}$ of solutions for (10) (deep discussion about Semenov's approximation see in [7]).

What we really did in the previous sections is to consider instead of (1) two types of Semenov's approximation:

1) $\mu_1 v = \lambda f(v)$ - for lower bound and
2) $\mu_2 v = \lambda f(v)$ - for upper bound

If both approximations have at least N intersections with $\lambda = \lambda_0$ such that between every two intersections of the second will be at least one intersection of the first, then our problem (1) will have at least N distinct solutions. In this case we even do not require that condition ii) for $f(x, \zeta)$ is satisfied, we only need condition iii) or iv). By choosing $f(\rho)$ with change small enough we can get constants $\tilde{C}_1$, $\tilde{C}_2$ closed to obtain the existence of N solutions for (1) with only information about behavior of $f_1(\rho)$ and $f_2(\rho)$.

Let us prove in this way the following

THEOREM 3. For any problem of the type (1) and any fixed $\lambda = \lambda_0 > 0$ there exists function $f_N(\zeta)$ such that there are at least N distinct positive solutions for (1) for $\lambda = \lambda_0$.

PROOF. It is enough to construct f on small interval $(\rho_1\ \rho_2)$ such that the problem (1) will have at least one solution on $(\rho_1\ \rho_2)$ and after this we can construct number of such an intervals as many as we wish.

From (9) it follows that

$$C_2 f^*(\rho) \leq \lambda \leq C_1 f^*(\rho) \tag{11}$$

where

$$f^*(\rho) = \frac{\rho}{f(\rho)e(x_0)}$$

Let (ρ_1^1, ρ_2^2) will be interval where $f(\rho) = \text{const.}$, or $f^*(\rho) = \gamma\rho$ and $f^*(\rho)$ crosses line $\lambda = \lambda_0$. By choosing $(\rho_1\ \rho_2)$ such that $\rho_1 < \rho^1$, $\rho_2 > \rho^2$ and:

1) $f(\rho)$ has local maximum at ρ^2 and local minimum at ρ^1 on the interval (ρ_1, ρ_2).

2) $C_1 = 1 + \delta$, $C_2 = 1 - \delta$ in (11) for δ sufficiently small.

After this we can construct N such an intervals $(\rho^1_{(i)}, \rho^2_{(i)})$ $i = 1,\ldots,N$.
On every $(\rho^1_{(i)}, \rho^2_{(i)})$ $f(\rho)$ is constant and on $(\rho^2_{(i)}, \rho^1_{(i+1)})$ is increased for example, as ρ^2. On every $(\rho^1_{(i)}, \rho^2_{(i)})$ $f^*(\rho)$ is $\gamma\rho$ and if crosses line $\lambda = \lambda_0$ on every interval; on intervals $(\rho^2_{(i)}, \rho^1_{(i+1)})$ $f^*(\rho)$ decreases for example, as $1/\rho$.

Let us note that in this case all conditions on $f(\rho)$ fullfilled even ii) with possibly very big const K, that depends on λ_0, N.

From continuity argument on the intervals $(\rho^2_{(i)}\ \rho^1_{(i+1)})$ if is easy to see that the problem (1) will have also at least one solution on every such interval and in the end we will have at least, 2N distinct positive solutions for (1) on the interval $(\rho^1_{(1)}\ \rho^{(2)}_{(N)})$.

The next question is the following: Can we construct $f(\rho)$ such that the problem (1) will have a countable number of distinct positive solutions? There are examples of this type (Gelfand's equation) but with nonlinearities that don't satisfy our conditions -particularly ii)- I expect that in our situation there are no more than finite number of solution (without bounds however), but I am unable to prove it.

BIBLIOGRAPHY

1. Amann, H., "Fixed point equations and nonlinear eigenvalue problem in ordered Banach spaces, SIAM, Rev. 18, 620-709 (1976).

2. Bony, J.M., "Principe du maximum dans les espaces de Sobolev". C.R. Acad. Sc. Paris, v. 265, ser A, 333-336, (1967).

3. Gidas, B., Ni. W.-H., and Nirenberg, L.,"Symetry and related properties via the maximum principle". Comm. Math. Phys. 68, 209-243, (1979).

4. Lions, P.L.,"On the existence of posi tive solutions of semilinear elliptic equations", SIAM Rev. 24, 441-467, (1982).

5. Rabinowitz, P.H., "Some global results for nonlinear eigenvalue problems", Journal of Functional Anal. 7, 487-513 (1971).

6. Schuchman, V., "About uniquenedd for nonlinear boundary calue problems", Math. Annalen. 267, pp 537-542 (1984).

7. Wiebers, H., "Dissertation", Hamburg, 1984.

DEPARTMENT OF MATHEMATICS
CENTRO DE INVESTIGACION Y DE
ESTUDIOS AVANZADOS DEL I.P.N.
APARTADO POSTAL 14-740
MEXICO, D.F. 07000
MEXICO.

Contemporary Mathematics
Volume 58, Part III, 1987

THE INDEX OF A VECTOR FIELD ON A COMPLEX ANALYTIC SURFACE WITH SINGULARITIES

José Seade

A classical result of Poincaré-Hopf tells us that if M is a smooth, closed n-manifold, then the number of zeroes of a vector field X on M with finite singularities, counted with their multiplicities, is independent of the vector field X and equals the Euler characteristic of M. In this work we study the following question: What happens if M is no longer smooth, but it has a finite number of singularities?. We restrict ourselves to dimension 4, although some of the definitions and results that we give here can be easily adapted to higher dimensions. This will be included in [3], where we will study this problem in relation with algebraic invariats associated to a holomorphic vector field. The problem is indeed of a local nature: If M is a variety, singular at the points $P_1,\dots,P_k$, by a vector field on M we mean a vector field X on the regular points of M. Now, given M as above and a vector field X on M, with finite singularities, what is the index of X at a singular point of M?. In this work we propose a definition for this number in the case when M has the structure of a complex 2-dimensional variety and its singularities are of a certain type: normal, Gorenstein and smoothable. (See the definitions in §1 below). In particular, every hypersurface or complete intersection germ satisfies these hypothesis, but many other interesting singularities are also included (as for example some of Dolgachev's triangle singularities [1,15], which are always normal and Gorenstein). This local index turns out to be canonical, in the sense that it depends only on the vector field X and on the complex structure on M around its singular points $P_1,\dots,P_k$. Using this local index we may define the total index of X on M in the usual way, and we prove in §1 the following theorem.

THEOREM. *Let M be a closed, complex 2-dimensional variety with finite singularities $P_1,\dots,P_k \in M$, and assume all of these singularities are normal, Gorenstein and smoothable. Then, the total index of a vector field on M with finite singularities is independent of the choice of the vector field and*

$$\begin{aligned}\mathrm{Ind}\,(X) &= \chi(M') \\ &= \chi(M) + \sum_{i=1}^{k} \mu_i\end{aligned}$$

© 1987 American Mathematical Society
0271-4132/87 $1.00 + $.25 per page

where M' *is a smoothing of* M *and* μ_i *is the Milnor number of* P_i, $i = 1,\dots,k$.

In the real case, when M is just differentiable, things are more complicated and there does not seem to be any natural way of defining the local index of a vector field X on M at a singularity of M. If the singularities of M have a Milnor number, in the sense of [23], then one can use Wall's result to get a $\mathbb{Z}/2$-index. In the case when the singularities of M can actually be endowed with a complex structure for which the singularities satisfy the hypothesis of our theorem above, we can again argue as in the complex case and obtain a similar result. However, this definition of the index does depend on the choice of the complex structure (because the Milnor number does, see 2.1 below) , so we can only obtain a $\mathbb{Z}/12$-invariant intrinsically associated to the vector field X and the real structure on M. By introducing in §2 an Arf invariant, defined by Rochlin in [17] and studied by the author in [19,21] in relation with complex singularities, we can improve this $\mathbb{Z}/12$-invariant by a factor of 2, thus obtaining a $\mathbb{Z}/24$-index intrinsically associated to X and the real structure on M, and which satisfies a result (2.2) similar to the theorem above modulo 24.

To finish this introduction, I would like to express my gratitude to Alberto Verjovsky for his continuous, stimulating influence and for many thought provoking discussions.

§1. A complex analytic surface V is *normal* at a point $P \in V$ if either P is a regular point of V or else P is an isolated singularity of V and every weakly holomorphic function at P is actually holomorphic [12]. A normal singularity $P \in V$ is *Gorenstein* [2,5] if there exists a nowhere vanishing holomorphic 2-form on a punctured neighbourhood of P in V. A normal singularity $P \in V$ is *smoothable* if there exists a deformation of P where all the nearby fibers are smooth. More precisely [2,4,6], P is smoothable if there exists the germ of a complex analytic variety W of dimension 3, such that $V \subset W$ and there exists a holomorphic map

$$F : W \to \Delta \subset \mathbb{C}$$

onto the germ of $\mathbb{C}$ at 0, such that V is $F^{-1}(0)$, $F^{-1}(t) = V_t$ is non-singular for $t \neq 0$, and F is flat at P, i.e. it is not a zero divisor in $\mathcal{O}_{W,P}$. Under these circumstances, W is either regular at P or else it has an isolated normal singularity there. If W is regular at P, then its germ at P is that of $\mathbb{C}^3$ at 0 and we are in the situation envisaged in the example (1.2) below. In any case, we may think of W as embedded in some complex space $\mathbb{C}^N$ and, following [10,2, or 4], consider a sufficiently small sphere S_ϵ in $\mathbb{C}^N$ around P so that $K = V \cap S_{\epsilon'}$ is the link of P in V for all $\epsilon' \leqslant \epsilon$, and we choose a sufficiently small disc $\Delta' \subset \Delta$ around 0, so that every fiber $F^{-1}(t)$, $t \in \Delta'$, meets the sphere S_ϵ transversally. Just as in [4], we call the intersection $V_t = \mathcal{V}_t \cap D_\epsilon$, $t \in \Delta'$, a *Milnor fibre* of P; V_t is an oriented 4-manifold with boundary diffeomorphic to the link K, and the diffeomorphism type of V_t is independent of the embedding of $\mathcal{W}$ in $\mathbb{C}^N$ and of the choices of ϵ,t. (Because $F|F^{-1}(\Delta'-\{0\}$ is a fibre bundle over $\Delta'-\{0\}$, by *Ehresmahn's* fribration lemma). We denote by F a fixed Milnor fibre of P. We note however that F does depend on the choice of the smoothing of P, generally speaking. (It does not for hypersurface or complete intersection germs). In all cases, F is always a parallelizable manifold [21] with the homotopy type of a CW-complex of dimension 2, its 1st-Betti number vanishes [6], and the number,

$$\mu = \text{Rank } H_2(V_t; \mathbb{Z})$$

called the *Milnor number* of P, is independent of the smoothing, by the formula of Laufer-Steenbrink [11,25]. Hence, the Euler characteristic of F is $\mu + 1$, independently of the smoothing.

We now wish to define the (local) index of a vector field X on $\mathcal{V}$, singular only at P. For this we recall that if P is a regular point of $\mathcal{V}$, then the index of X at P is defined [9] by identifying the germ of $\mathcal{V}$ at P with that of $\mathbb{R}^4$ at 0, and taking the degree of the map $\frac{X}{|X|}: S_\epsilon \to S^3$, where S_ϵ is a small sphere around 0. In the singular case, our definition of the (local) index of X at P is similar: Since the Milnor fibre F of P is parallelizable, F can be immersed in $\mathbb{R}^4$, by the theorem of Hirsch-Poénaru [16]. The vector field X on V-{0} can be moved by an isotopy and considered as a vector field on the boundary of F, which we identify with the link K. If $g: F \to \mathbb{R}^4$ is an immersion, we have a well defined map (up to homotopy)

$$g(X): K \to S^3 \subset \mathbb{R}^4$$

given by $Z \to g_*(X(Z))$, where g_* is the derivative of g at a point $Z \in K$. Then, $g(X)$ represents a homotopy class in $[K,S^3] \cong \pi_3(S^3) \cong \mathbb{Z}$ and therefore it has a degree. We claim (1.3 below) that the degree of $g(X)$, denoted Ind(X), is independent of all choices, depending only on the complex structure at P and on the vector field X.

1. DEFINITION. The number $\text{Ind}(X) = \text{Ind}(X,P) \in \mathbb{Z}$ is the *(local) index* of X at $P \in \mathcal{V}$.

It is clear that this definition reduces to the usual one when P is a regular point of M.

2. EXAMPLE. Let $f: (\mathbb{C}^3,0) \to (\mathbb{C},0)$ be (the germ of) a holomorphic map with an isolated critical point at $0 \in \mathbb{C}^3$, and let $\mathcal{V} = f^{-1}(0)$. Then, $\mathcal{V}$ is a complex analytic surface with a normal Gorenstein singularity at $0 \in \mathbb{C}^3$, and this

singularity is smoothable, for the map f itself gives the smoothing. (The fibers are the non-critical levels of f). If we let X be a vector field on V that is everywhere transversal to the link K, then

$$\mathrm{Ind}(X) = \mu + 1 ,$$

where μ is the Milnor number of 0, which in this case equals the multiplicity of 0 as critical point of f. To see this, we argue as in [20], lemma 1.2, and show that the vector field X on $K = \partial F$ can be extended to the whole of F minus $(\mu+1)$ small discs D_i, $i = 1,\dots,\mu+1$, in such a way that if we take the immersion of F to be an embedding on each D_i, then each boundary sphere $S_i = \partial D_i$ contributes with 1 for the index of X.

3. PROPOSITION. *The integer* $\mathrm{Ind}(X)$ *depends only on* X *and on* (V,P), *and not on the various choices involved in its definition.*

To prove this proposition we need to show that $\mathrm{Ind}(X)$ does not depend neither on the choice of the smoothing nor on the immersion of the Milnor fibre. Similar arguments to those of 1.2 above show that if X is everywhere transversal to the link K, then its index equals the Euler characteristic of the Milnor fibre F, independently of the choice of the immersion, and we also know that $\chi(F) = \mu + 1$ is independent of the smoothing, so the proposition is proved when X is everywhere transversal to the link K. To complete the proof of the proposition we reduce the general case to that when X is transversal to the link by observing that if X and Y are continuous vector fields on V, singular only at P, then at each point x of $V-\{P\}$ the vectors $X(x)$ and $Y(x)$ differ, up to homotopy, by an element in $SO(4)$ that factors through $SO(3)$. Hence, X and Y differ by a continuous map

$$d = d(X,Y)\colon K \to SO(4)/SO(3) \cong S^3$$

and the degree of d is the difference $\mathrm{Ind}(X) - \mathrm{Ind}(Y)$. Therefore $(\mathrm{Ind}(X)-\mathrm{Ind}(Y))$ depends only on X and Y, and the result follows.

We now let M be a closed, connected, complex analytic surface with isolated singularities $P_1,\dots,P_k$, which are all normal, Gorenstein and smoothable. If X is a continuous vector field on M singular at the P_i's and, possibly, at a finite number of regular points of M, then its *total index*, denoted by $\mathrm{Ind}(X)$, is the sum of all the local indices of X.

4. THEOREM. *The total index of a vector field* X *on* M *with finite singularities is independent of the vector field and*

$$\mathrm{Ind}(X) = \chi(M) + \sum_{i=1}^{k} \mu_i$$

where μ_i *is the Milnor number of* P_i, $i = 1,\dots,k$.

Proof: Let M' be a closed 4-manifold obtained by attaching to $M-\{P_1,\ldots,P_k\}$ a Milnor fibre of each P_i, and let $F_1,\ldots,F_k$ be these fibers. We observe that, essentially by definition, the local index of X at a point P_i can be identified with the obstruction for extending X to a vector field on the fibre F_i. Hence $\operatorname{Ind}(X) = \chi(M')$, by the theorem of Poincaré-Hopf stated in the introduction. The result now follows because

$$\chi(M') = \chi(M) + \sum_{i=1}^{k} \mu_i$$

since $\chi(F_i) = \mu_i + 1$ and the Euler characteristic of any oriented 3-manifold is 0.

§2. We now let M be a connected, closed 4-dimensional real variety, smooth and orientable away from a finite number of points $P_1,\ldots,P_k$. We suppose further that the germ of M at each P_i is diffeomorphic to the germ of a complex analytic surface V_i at a normal, Gorenstein, smoothable singularity P'_i.

If X is a continuous vector field on M, then an attempt to define the local index of X at a P_i could be by endowing the germ of M at P_i with a complex structure as above and then defining $\operatorname{Ind}(X,P_i)$ as in §1. This definition is fine and yields a result similar to 1.4 above, once we have fixed the complex structure around each P_i. However, this definition does depend on the choice of the complex structure around the P_i's, as is shown in the following example.

1. EXAMPLE. Consider the singularities at 0 defined by the complex polynomials in $\mathbb{C}^3$,

$$f_1(Z) = Z_1^2 + Z_2^7 + Z_3^{14}; \quad f_2(Z) = Z_1^3 + Z_2^4 + Z_3^{12}$$

By [16 or 14], the links K_1, K_2 of these two singularities are orientation preserving diffeomorphic, since both fiber as S^1-bundles with Euler number -1 over a Riemann surface of genus 3. However, the index at 0 of the unit outwards normal field of K_1 in $f_1^{-1}(0)$ is 79, while that on K_2 has index 67.

Hence, the above definition of the (local) index does not yield something intrinsec of the real structure on M. Neverthless, that definition of the index does yield a $\mathbb{Z}/12$-invariant of X, as we show below (2.2), and by adding to it a $\mathbb{Z}/2$-correction term we can actually get a $\mathbb{Z}/24$-invariant of X, which depends only on the real structure on M and yields to a result similar to the theorem above, but only mod(24).

For this, we recall [14] that if (V,P) denotes the germ of a complex analytic normal surface singularity, then a *resolution* of P consists of a non-singular surface $\widetilde{V}$, together with a projection map $\pi: \widetilde{V} \to V$, which is holomorphic and such that $\pi^{-1}(V-\{P\})$ is dense in V and π is a biholomorphism there. We now observe that if $P \in V$ is Gorenstein, then a nowhere-zero holomorphic 2-form on $V-\{P\}$ defines as in [20,21] a reduction to $SU(2)$ of the structure group of the tangent bundle TV of $V-\{P\}$; therefore [21], a vector field X on $V-\{P\}$ defines automatically a continuous trivialization ρ_X of the complex vector bundle TV. This trivialization of TV can be lifted to a trivialization ρ_X of the tangent bundle of $\tilde{V}-\pi^{-1}(P)$. If $\tilde{V} \subset \tilde{V}$ denotes a compact tubular neighbourhood of the exceptional curve $\pi^{-1}(P)$ in $\tilde{V}$, then [7] the lifting $\rho_{\tilde{X}}$ of ρ_X yields representatives $C_1(X)$, $C_2(X)$ of the Chern classes of $\tilde{V}$ that vanish over its boundary $\partial\tilde{V} \cong K$; they live in $H^*(\tilde{V}, \partial\tilde{V}; \mathbb{Z})$. The 2nd Chern class $C_2(X)$ is the obstruction for extending to all of $\tilde{V}$ the lifting $\tilde{X}$ of X. The 1st Chern class $C_1(X)$ turns out to be independent of the choice of X, see [2 or 19], its image in $H^2(\tilde{V}; \mathbb{Z})$ is the usual Chern class of $\tilde{V}$, and $C_1(X)$ is the Poincaré-Lefschetz dual of the *canonical class* $-K \in H_2(\tilde{V}; \mathbb{Z})$; the class K is characterized for being the unique homology class that satisfies the adjunction formula

$$\partial g - 2 = S \cdot (S + K)$$

for every compact Riemann surface S of genus g embedded in the interior of $\tilde{V}$, and the dot denotes the intersection of cycles [6]. The class K may indeed be always represented by an oriented real submanifold of $\tilde{V}$ (see [19; p.5]), which is a characteristic submanifold of $\tilde{V}$, in the sense of [19] (see also [17, 13]). Thus K has an associated Rochlin-Arf invariant $\text{Arf } K \in \mathbb{Z}_2$, which is the Arf-invariant of a certain quadratic form $q \in H_1(K; \mathbb{Z}_2)$ introduced by Rochlin in [17]. (We refer to the appendix in [18] for a clear account on Arf invariants).

We remark that if P is smoothable and F is a Milnor fibre of P, then the above trivialization ρ_X of TV also defines a trvialization $\rho_{X'}$ of $TF|\partial F$, which yields relative Chern classes $C_i(X)$, $i = 1,2$, in $H^*(F, \partial F; \mathbb{Z})$. The 1st Chern class $C_1(X)$ always vanishes, by [21; 3.2], and the value of $C_2(X)$ on the orientation cycle of F is the local index $\text{Ind}(X,P)$ of §1.

Now, if M is as before and $P \in M$ is a singularity, whose germ can be endowed with two different complex structures (V,P), (V',P) for which P is normal, Gorenstein and smoothable, then

$$\mu - 12 \text{ Arf } K \equiv \mu' - 12 \text{ Arf } K' \qquad \text{mod}(24)$$

where the μ's and K's are the corresponding Milnor numbers and canonical classes, by [21; 4.5]. Hence, if X is a vector field on M, then the number

$$I(X,P) = \text{Ind}(X,P) - 12 \text{ Arf } K \qquad \text{mod}(24)$$

is independent of the complex structure, where $\text{Ind}(X,P)$ denotes the index of X at P as defined in §1. We call $I(X,P) \in \mathbb{Z}/24$ *the local index* of X at P.

We remark that Arf K depends only on the choice of the complex structure at P and not on the vector field X. Indeed, if $g(P)$ is the geometric genus of P, see [8], then Arf K is the mod(2)-reduction of $g(P)$, by [21; 4.1], because the conjecture of Laufer [8] mentioned there was proved by Steenbrink in [22].

The *total index* of X on M is now defined to be the mod(24)-reduction of the sum of all the local indices of X. We denote this index by $I(X) \in \mathbb{Z}/24$.

2. THEOREM. *Let* M *be a closed, real variety of dimension 4 as before, singular at the points* $P_1,\ldots,P_k$. *Then the total index of a field* X *on* M *with finite singularities is independent of the choice of the vector field and*

$$I(X) = \chi(M) + \sum_{i=1}^{k} (\mu_i - 12 \text{ Arf } K_i) \qquad \text{mod } 24$$

$$= \chi(M) + \sum_{i=1}^{k} (\mu_i - 12\, g(P_i)) \qquad \text{mod } 24$$

where μ_i, $i = 1,\ldots,k$, *is the Milnor number of* P_i *for some complex structure for which* P_i *is normal, Gorenstein and smoothable,* $g(P_i)$ *is the geometric genus of* P_i *with respect to that complex structure, and* Arf $K_i \in \{0,1\}$ *is the Rochlin--Arf invariant [17] of a non-singular representative of the canonical class* K_i *of some resolution of* P_i.

The proof of this result is easy and analogous to that of 1.4 above.

BIBLIOGRAPHY

[1] I. V. Dolgachev. *Quotient conical singularities of complex surfaces.* Funct. Anal.Appl. 8(1984), 160-167.

[2] A. Durfee. *The signature of smoothings of complex surface singularities.* Math.Ann. 232(1978), 85-98.

[3] X. Gômez-Mont, J. A. Seade and A. Verjovsky. On the Milnor number of a singular point. Preprint.

[4] G.-M. Greuel and J. Steenbrink. *On the topology of smoothable singularities.* AMS Proc.Symp. Pure Maths. vol. 40 , part 1, (1983), 85-97.

[5] R. Hartshorne. *Residues and Duality.* Lecture Notes in Maths. 20(1966), Springer-Verlag.

[6] R. Hartshorne. *Algebraic Geometry.* Graduate texts in Maths. 52, Springer--Verlag.

[7] M. Kervaire. *Relative characteristic classes.* Am. J. Maths. 79(1957), 517-558.

[8] H. B. Laufer. *On μ for surface singularities.* AMS Proc.Symp. Pure Maths. 30, part 1 (1977), 45-49.

[9] J. Milnor. *Topology from the differentiable viewpoint.* University Press of Virginia, 1965.

[10] J. Milnor. *Singular points of complex hypersurfaces.* Ann. of Maths. Study 61(1968), Princeton Univ. Press.

[11] J. Milnor. *On the 3-dimensional Brieskorn manifolds* M(p,q,r). In Knots, groups and 3-manifolds (L.P. Neuwirth, Ed.), Ann. of Maths. Study 84(1975), 175-225.

[12] R. Narasimham. *Introduction to the theory of analytic spaces.* Lecture Notes in Maths. 25(1966), Springer-Verlag.

[13] S. Ochanine. *Signature modulo 16, invariantes de Kervaire généralisés et nombre characteristic dans la K-théorie réelle.* Mem.Soc.Math. de France, 5(1981).

[14] P. Orlik and Ph. Wagreich. *Isolated singularities of algebraic surfaces with* $\mathbb{C}^*$*-action.* Ann. of Maths. 93(1971), 205-228.

[15] H. Pinkham. *Normal surface singularities with* $\mathbb{C}^*$*-action.* Math.Ann. 227 (1977), 183-193.

[16] V. Poênaru. *Sur la théorie des immersions.* Topology 1(1962), 81-100.

[17] V.A. Rochlin. *Proof of Gudkov's hypothesis.* Funct. Anal. Appl. 6(1972), 136-138.

[18] C.P. Rourke and D. Sullivan. *On the kervaire obstruction.* Ann. of Maths. 94(1971), 397-413.

[19] J.A. Seade. *Singular points of complex surfaces and homotopy.* Topology 21(1982), 1-8.

[20] J.A. Seade and B.F. Steer. *The elements of* π_3^S *represented by invariant framings on quotients of* $\widetilde{SL}_2(\mathbb{R})$ *by certain discrete subgroups.* Adv. in Maths. 46(1982), 221-229.

[21] J.A. Seade. *A cobordism invariant for surface singularities.* AMS. Proc. Symp. Pure Maths. 40, part 2 (1983), 479-484.

[22] J. Steenbrink. *Mixed Hodge structures associated with isolated singularities.* AMS Proc. Symp. Pure Maths. 40, part 2(1983), 513-536.

[23] C.T.C. Wall. *Topological invariance of the Milnor number mod 2.* Topology 22(1983), 345-350.

INSTITUTO DE MATEMATICAS
Universidad Nacional Autónoma de Mexico
México, 04510, D.F.
MEXICO

Contemporary Mathematics
Volume **58**, Part III, 1987

A UNIFORMIZATION THEOREM FOR HOLOMORPHIC FOLIATIONS

Alberto Verjovsky

INTRODUCTION. One of the fundamental results in the theory of functions of one complex variable is the Uniformization Theorem of Koebe and Poincaré. This theorem does for Riemann surfaces what Riemann's mapping theorem does for plane regions. Given a complex manifold M with a holomorphic nonsingular foliation F, by Riemann surfaces, one is led to the following question: Can we uniformize the leaves of F in a continuous or differentiable fashion? Put more simply the question is: Does a "foliated" uniformization theorem exist? We prove a uniformization theorem for hyperbolic foliations.

One natural condition to impose on M is to suppose that M is a complete (but not necessarily compact) hyperbolic manifold in the sense of Kobayashi [K]. This condition implies, in particular, that every leaf $L \in F$ can be uniformized with the unit disk $\Delta = \{z \in \mathbb{C}\,;\, |z|<1\}$. Such foliations will be called *completely hyperbolic foliations*. In general, for an arbitrary manifold M with a holomorphic foliation F by Riemann surfaces one could have leaves of mixed type: leaves that are uniformized by the complex plane $\mathbb{C}$ (*parabolic leaves*) and leaves that are uniformized by the disk Δ (hyperbolic leaves). The case when there exists a leaf L which is uniformized by the Riemann sphere $\hat{\mathbb{C}} = \mathbb{C} \cup \{\infty\} = \mathbb{C}P^1$ is settled by the following theorem due to Reeb:

(1.1) THEOREM (Reeb stability theorem). *Let* M *be a complex compact manifold and let* $\dim_{\mathbb{C}}(M) = n$. *Suppose* F *is a nonsingular holomorphic one-dimensional foliation such that one leaf is uniformized by* $\hat{\mathbb{C}}$. *Then every leaf is also uniformized by* $\hat{\mathbb{C}}$. *The leafspace* M/F *is a* $(n-1)$-*dimensional, compact, complex Hausdorff manifold and the natural projection* $\pi : M \to M/F$ *is a locally-trivial holomorpic fibration.*

Foliations by Riemann surfaces appear naturally when we integrate nonsingular holomorpic vector fields and consider its solutions, $\phi(T)$, as functions of a complex parameter $T \in \mathbb{C}$. The existence of a nonsingular holomorphic vector field on a compact manifold, M, imposes severe restrictions on the manifold. Since the vector field is complete, every leaf is parabolic; in fact, every leaf of the induced foliation must be analytically equivalent to the quotient of $\mathbb{C}$ by a discrete subgroup of translations. Hence, M cannot be

© 1987 American Mathematical Society
0271-4132/87 $1.00 + $.25 per page

hyperbolic. Also, the Euler-Poincaré characteristic of M must vanish. To make a better analysis one should also consider singular (or even meromorphic) vector fields on complex manifolds. Especial emphasis should be put in the local study of solutions of a holomorphic vector field near an isolated singularity (having just isolated singularities is a local generic condition). In this paper we also deal with this local analysis.

§2 HOLOMORPHIC FOLIATIONS AND THE KOBAYASHI METRIC. Let M be a complex manifold of complex dimension n.

(2.1) DEFINITON. A *holomorphic foliation*, F, on M of codimension q, $0 \leqslant q \leqslant n$, consists of a maximal holomorphic atlas

$$A = \{(U_i, \phi_i);\ \phi_i : U_i \to \mathbb{C}^n\}_{i \in I}$$

such that the transition functions

$$\phi_{ij} = \phi_i \circ \phi_j^{-1} : \phi_j(U_i \cap U_j) \to \phi_i(U_i \cap U_j)$$

are of the form

$$\phi_{ij}(x,y) = (\alpha_{ij}(x,y),\ \gamma_{ij}(y));\ x \in \mathbb{C}^{n-q},\ y \in \mathbb{C}^q .$$

Since the transition functions send a subset of the affine subspace $y =$ constant into a similar set, it follows that M is partitioned as a disjoint union of connected, holomorphically immersed submanifolds of codimension q. Such submanifolds are called the *leaves* of the foliation and if L is a leaf, we write $L \in F$. The union of the tangent spaces of the leaves is a $(n-q)$-distribution, or a holomorphic subbundle, TF, of the tangent bundle of M.

Alternatively, a holomorphic foliation is obtained when one integrates an involutive holomorphic subbundle $F \subset TM$.

The case $q = n-1$ is the one in which we are interested. Every holomorphic line bundle contained in the tangent bundle of M is tangent to a foliation by Riemann surfaces (i.e., every holomorphic line bundle is integrable). Local integrability is a consequence of the existence and regularity theorems of solutions of holomorphic vector fields. Therefore, there exists a one-to-one correspondence between holomorphic one-dimensional foliations and holomorphic tangent line bundles of M. These line bundles are holomorphic sections $\sigma : M \to G_1(M)$, where $G_1(M)$ is the projectivization of the tangent bundle of M. In general the space of such sections is compact and this accounts for the rigidity of holomorphic nonsingular foliations.

(2.2) DEFINITION. Let M be a complex manifold and let F be a one-dimensional holomorphic foliation. A leaf $L \in F$ is called *parabolic* if there exists a holomorphic covering map $\pi: \mathbb{C} \to L$. A leaf L is called *hyperbolic* if it is holomorphically covered by the unit disk Δ. A leaf L is called *elliptic* if there exists a holomorphic nonconstant map $\pi: \hat{\mathbb{C}} \to L$.

Because of Reeb stability (1.1), we will only consider one-dimensional holomorphic foliations without elliptic leaves.

(2.3) DEFINITION. Let M be a complex, connected manifold. The *Kobayashi pseudo-distance*, d_M, on M is defined as follows. Let $\rho: \Delta \times \Delta \to \mathbf{R}$ denote the Poincaré distance obtained from the conformal metric of constant negative curvature on the unit disk. Given points $x, y \in M$, we choose a finite chain of points,

$$x = x_0, \ldots, x_{k-1}, x_k = y, \text{ of } M, \text{and points } a_1, \ldots, a_k, b_1, \ldots, b_k$$

of the unit disk Δ and holomorphic mappings f_i, $1 \leq i \leq k$, such that the following condition is satisfied:

$$(*) \quad \begin{cases} f_i: \Delta \to M \\ f_i(a_i) = x_{i-1} \quad \text{and} \quad f_i(b_i) = x_i . \end{cases}$$

For each choice of points and mappings satisfying (*), define $\rho_*(x,y)$ by

$$\rho_*(x,y) = \rho(a_1, b_1) + \ldots + \rho(a_k, b_k) .$$

Finally, define $d_M(x,y)$ as the infimum of $\rho_*(x,y)$ for all possible choices of points and mappings satisfying (*).

The function $d_M: M \times M \to \mathbf{R}$ is continuous and satisfies the properties of a pseudo-distance. It could be identically zero, for example if $M = \mathbb{C}^n$.

We recall some properties of the Kobayashi pseudo-distance:

i) When $M = \Delta$, the unit disk, $d_M = \rho =$ Poincaré distance. More generally if $M = B^n$, the open unit ball in $\mathbb{C}^n$, then d_M is the Poincaré-Bergman metric, hence is a Kaehler metric.

ii) If $f: M \to N$ is a holomorphic mapping, then:

$$d_N(f(x), f(y)) \leq d_M(x,y) , \text{ for } x, y \in M .$$

Therefore holomorphic mappings are distance-decreasing. In particular if f is a biholomorphism then $d_M(x,y) = d_N(f(x), f(y))$.

iii) Let M be a complex manifold and let d' be any pseudo-distance such that

$$d'(f(a), f(b)) \leq \rho(a,b), \text{ for } a, b \in \Delta$$

for any holomorphic mapping $f: \Delta \to M$. Then

$$d_M(x,y) \geq d'(x,y) , \text{ for } x, y \in M .$$

Hence, d_M is the largest pseudo-distance for which every holomorphic mapping $f:\Delta \to M$ is distance-decreasing.

It will be more convenient for us to define the Kobayashi distance in terms of a Finsler structure as follows:

Let M be a complex manifold. For each $x \in M$ let $F(x) \subset T_xM$ be defined by,

$$F(x) = \{\dot{\gamma}(0) = [\frac{d}{dT}\gamma]_{T=0} \ ; \ \gamma \in H_x\}$$

where H_x is the set of holomorphic mappings $h:\Delta \to M$ such that $h(0) = x$.

Let $K_M(x) = K(x)$ be the convex envelope of $F(x)$ in T_xM. Then this convex set is absorbent and balanced with respect to complex scalars. Therefore $K(x)$ determines, via the Minkowski functional, a complex semi-norm $\|\cdot\|_x : T_xM \to \mathbb{R}$.

The family of convex sets $\{K(x); x \in M\}$ varies continuously with respect to x, so that the function $x \to \| \ \|_x$ is continuous.

This defines the *Kobayashi pseudo-Finsler* structure

$$\|\cdot\| : TM \to \mathbb{R} .$$

(2.4) REMARK. A holomorphic mapping $f:M \to N$ satisfies

(**) $f(K(x)) \subset K(f(x))$ for all $x \in M$,

and equality of these convex sets, for every $x \in M$, occurs if and only if f is a biholomorphism.

Property (**) is Kobayashi's generalized Schwarz lemma [K].

(2.5) DEFINITION. For a complex manifold M define, for each $x, y \in M$:

$$d_M(x,y) = \inf_{\gamma \in \Omega(x,y)} \{\int \|\dot{\gamma}(t)\| \, dt\}$$

where $\Omega(x,y)$ denotes the space of all curves which are piecewise differentiable and join x with y. This is an equivalent definition of the Kobayashi pseudo-metric.

(2.6) REMARKS. 1) The convex sets $K(x)$ could be unbounded and even equal to all of T_xM. For example $M = \mathbb{C}^n$.

2) For the n-dimensional disk $B^n = \{(z_1,\ldots,z_n) \in \mathbb{C}^n \ ; \ \sum|z_i|^2 < 1\}$ the Kobayashi metric is Riemannian and Kähler, since it coincides with the Poincaré-Bergman metric. In general it is only continuous and not differentiable for bounded open sets in $\mathbb{C}^n$.

(2.7) DEFINITION. A complex manifold M is said to be *hyperbolic* (in the sence of Kobayashi) if the convex sets $K(x)$ are bounded for all $x \in M$. This is equivalent to the fact that the Kobayashi pseudo-metric is actually a metric. A hyperbolic manifold is said to be a *complete hyperbolic manifold* if its

Kobayashi metric is complete.

(2.8) EXAMPLES. 1) Any bounded domain $\Omega \subset \mathbb{C}^n$ is hyperbolic and any analytic polyhedron is complete hyperbolic.

2) $\Delta^n = \Delta \times \ldots \times \Delta \subset \mathbb{C}^n$, the n-polydisk, is complete hyperbolic and

$$d_{\Delta^n}((x_1,\ldots,x_n);\ (y_1,\ldots,y_n)) = \operatorname{Max}\{\rho(x_i,y_i)\}.$$

Every biholomorphic mapping $f: \Delta^n \to \Delta^n$ is of the form $f(x_1,\ldots,x_n) = (g_1(x),\ldots,g_n(x_n))$, where g_i belongs to the group of biholomorphisms of Δ, $\operatorname{Aut}(\Delta) \cong PSL(2,\mathbb{R})$. Therefore $\operatorname{Aut}(\Delta^n) \simeq (PSL(2,\mathbb{R}))^n = PSL(2,\mathbb{R})\times\ldots\times PSL(2,\mathbb{R})$.

3) The finite product of complete hyperbolic manifolds is complete and hyperbolic. Therefore, since any Riemann surface of genus greater than one is complete hyperbolic, it follows that finite products of Riemann surfaces of genus greater than one are also comlete and hyperbolic. A complex manifold is complete hyperbolic if and only if its universal covering is complete and hyperbolic.

(2.9)) DEFINITION. Let M be a complex manifold. For each $x \in M$, let $P(x)$ be the maximal linear complex subspace of T_xM which is contained in $K(x)$. Let $H(x)$ be an orthogonal complement of $P(x)$, then:

$$K(x) = (H(x) \cap K(x)) + P(x)$$

where "+" indicates sum of convex sets. For every $x \in M$, $H(x) \cap K(x)$ is a bounded convex set. We call $\dim_{\mathbb{C}}(P(x)) := r(x)$, the *rank* at x of the complex manifold M. The *dimension of hyperbolicity at* x, denoted by $h(x)$ is defined by

$$h(x) = \dim_{\mathbb{C}} H(x) = \dim_{\mathbb{C}} M - r(x).$$

If $h(x) > 0$ for all $x \in M$ we say that M is a *semi-hyperbolic* manifold. The subspace $P(x) \subset T_xM$ is well-defined and it is called the *parabolic subspace* at x. Hyperbolicity is equivalent to the condition that $r(x) = 0$ for all $x \in M$.

(2.10) REMARK. If $f: M \to N$ is a biholomorphism then since $f(K(x)) = K(f(x))$ we see that $h(x) = h(f(x))$ and $r(x) = r(f(x))$. The functions r and h are in general only semicontinuous.

(2.11) EXAMPLES 1) Let $B^n \subset \mathbb{C}^n$ be the unit open ball in $\mathbb{C}^n$, $n>1$, and let $\pi:\tilde{B}^n \to B^n$ be the holomorphic map which corresponds to blowing up the origin. Then $\pi^{-1}(0)$, the exceptional divisor, is $\mathbb{C}P^{n-1}$ and therefore it has identically vanishing Kobayashi metric and at every point of $\tilde{B} - \pi^{-1}(0)$ the convex set $K(x)$ is bounded. This implies that $\tilde{B}^n - \pi^{-1}(0)$ is hyperbolic but not complete. Therefore, $B^n - \{0\}$ is an incomplete hyperbolic manifold.

2) Typical examples of semi-hyperbolic manifolds which admit hyperbolic foliations are the following: Let M_g be a compact Riemann surface of genus, g, greater than one and let $\Psi: \pi_1(M_g) \to \mathrm{Aut}(\hat{\mathbb{C}})$ be a representation of its fundamental group into the group of Möbius transformations. Let $\pi: \Delta \to M_g$ be a uniformization of M_g. Then, $\Gamma := \pi_1(M_g)$ acts biholomorphically on $\Delta \times \mathbb{C}^n$ as follows:

$$g(x,y) = (g(x), \Psi(g)(y)) ; \quad x \in \Delta , \quad y \in \hat{\mathbb{C}} , \quad g \in \Gamma ,$$

where we identify Γ with the group of deck transformations of the covering π

In this manner Γ acts freely and properly discontinuosly. The quotient $\Delta \times \hat{\mathbb{C}}/\Gamma$ obtained by the ortbits of the action is a compact, complex 2-dimensional manifold, denoted by $M(\Psi)$. The canonical projection $p: \Delta \times \hat{\mathbb{C}} \to M(\Psi)$ is a covering projection. The action of Γ on $\Delta \times \hat{\mathbb{C}}$ preserves the "horizontal" and "vertical" foliations whose leaves are, respectively, of the form $\Delta \times \{\text{point}\}$ and $\{\text{point}\} \times \hat{\mathbb{C}}$. These two foliations descend to $M(\Psi)$ to two foliations denoted by F_h and F_v, respectively. There exists a holomorphic fibration $f: M(\Psi) \to M_g$ whose fibres are the leaves of F_v, which are copies of $\hat{\mathbb{C}}$. The leaves of F_h are transverse to the fibres of f and the restriction of f to a horizontal leaf is a covering of M_g. If $(x,y) \in \Delta \times \hat{\mathbb{C}}$ then the mapping $\pi_y : \Delta \to L_{p(x,y)}$, from the unit disk to the horizontal leaf passing through $p(x,y)$, defined by $\pi_y(T) = p(\pi(T),y)$ is a uniformization of $L_{p(x,y)}$.

In this example we see that $M(\Psi)$ is a semi-hyperbolic manifold of constant rank. The family of parabolic subspaces of $M(\Psi)$ is precisely the line bundle tangent to the vertical fibres. This example raises a question:

(2.12) QUESTION. Let M be a complex manifold of constant rank. Is it true that the parobolic subspaces form a holomorphic subbundle of TM? If such is the case is such parabolic subbundle integrable? It is not difficult to show that under the hypothesis of constant rank the family $\{P(x); x \in M\}$ is a continuous subbundle of TM.

The construction given in example 2) works in general when one is given a representation $\Psi : \Pi_1(M) \to \mathrm{Aut}(N)$, form the fundamental group of the complex manifold M into the group Aut(N) of biholomorphisms of the complex manifold N. The associated horizontal foliation is said to be obtained by *suspension* of the representation Ψ. Reciprocally, given a locally-trivial holomorphic fibration $f: M^m \to N^n$ with M^n compact, and a holomorphic foliation F of dimension n such that the leaves are transverse to the fibres of f it follows that F is holomorphically equivalent to a foliation obtained by a suspension of a representation $\Psi: \Pi_1(N) \to \mathrm{Aut}\, F$ where F is the fibre of f. The representation Ψ is the global holonomy of the foliation.

If M_g is a Riemann surface of genus greater than one, there exist interesting representations of its fundamental group. For example, if $\Psi:\Pi_1(M_g) \to \mathrm{Aut}(\hat{\mathbb{C}})$ is a representation into a fuchsian group, then the horizontal foiation in $M(\Psi)$ has the property of every leaf has the same limit set and this common limit set is a differentiable circle bundle over M_g, where each fibre is a round circle in a vertical fibre. If the representaion is quasi-fuchsian, then the common limit set of the horizontal leaves is a topological circle bundle over M_g, where each fibre is a quasi-circle in a vertical fibre. The circle bundle is differentiable if and only if the representation is fuchsian. On the other hand, let M^3 be a compact real hyperbolic manifold of constant negative curvature equal to minus one. Suppose that M^3 fibres over the circle with fibre a surface, M_g, of genus greater than one. For each $g>1$, there exists an infinity of such manifolds constructed by Jørgensen and Thurston [J][Th]. Let $\Pi: \mathbb{H}^3 \to M^3$ be an isometric covering of M^3 by hyperbolic space $\mathbb{H}^3$. Let $\Psi: \Pi_1(M^3) \to PSL(2,\mathbb{C}) = \mathrm{Aut}\,(\hat{\mathbb{C}})$ be a representation of $\Pi_1(M^3)$ into the group of isometries of $\mathbb{H}^3$ (which is isomorphic to $PSL(2,\mathbb{C})$) obtained by the deck transformations of Π. The image of Ψ is a discrete subgroup $\Gamma \subset PSL(2,\mathbb{C})$ and we have an exact sequence

$$0 \to \Pi_1(M_g) \overset{\alpha}{\to} \Pi_1(M^3) \overset{\beta}{\to} \mathbb{Z} \to 0$$

which comes from the homotopy exact sequence of the fibration $f: M^3 \to S^1$.

Then, $\Psi \circ \alpha: \Pi_1(M_g) \to PSL(2,\mathbb{C})$ gives an injective representation from which we obtain a complex 2-manifold $M(\Psi \cdot \alpha)$, whose horizontal foliation is a minimal foliation by Riemann surfaces. Minimality is a consequence of the fact that M has bounded hyperbolic volume and, therefore, Γ as well as any normal subgroup of Γ acts minimaly on $\hat{\mathbb{C}}$.

Thus, we can associate to each Jørgensen-Thurston manifold a complex surface with a minimal foliation by hyperbolic Riemann surfaces.

Let $\Psi: \Pi_1(M_g) \to \mathrm{Aut}(\hat{\mathbb{C}})$ be any representation. Then the tangent bundle of $M(\Psi)$ splits as a Whitney sum

$$T(M(\Psi)) = T(F_h) \oplus T(F_v) .$$

One can choose a hermitian metric $<,>$ on $M(\Psi)$ for which every horizontal leaf has constant negative curvature and every vertical leaf has positive constant curvature. Let $N \subset T(M(\Psi))$ be the unit sphere bundle of the line bundle tangent to the horizontal foliation. Then N is a 5-dimensional manifold and it is a circle bundle over $M(\Psi)$. Let $p: N \to M(\Psi)$ be the S^1-fibration. We have three smooth flows $g_t: N \to N$, $h_t^+ : N \to N$ $h_t^- : N \to N$ called respectively, the *foliated geodesic flow, stable foliated horocyclic flow and unstable foliated horocyclic flow.*

These three flows are defined on each leaf in the usual manner [A - A].

Let $\Psi : \Pi_1(M_g) \to \mathrm{Aut}(\hat{\mathbb{C}})$ be a representation obtained from a Jørgensen-Thurston compact hyperbolic manifold. The following proposition is not difficult to prove using the fact that both the horizontal foliation and the stable and unstable horocyclic flows on a compact surface of constant negative curvature are minimal.

(2.13) PROPOSITION. The flows h_t^+ and h_t^- are both minimal flows.

3) If M is a complex surface such that there exists a locally-trivial holomorphic fibration $\Pi: M \to M_g$, over a compact surface of genus $g > 1$ and whose fibre is a Riemann surface of genus $g' > 1$, then any foliation by Riemann surfaces transverse to the fibre must have finite global holonomy and every leaf must be compact. This follows from Hurwitz theorem that says that $M_{g'}$ has finite automorphism group of order at most $84(g'-1)$. When $g' = 1$, then the fibre is an elliptic curve $\mathbb{C}/\Lambda$, where Λ is a lattice. If $\mathbb{C}/\Lambda$ admits no automorphism as a complex Lie group, then its group of biholomorphisms consists only of translations and it is therefore abelian. In this case all representations are of the form given by the following composition:

$$\Psi:\Pi_1(M_g) \to H_1(M_g, \mathbb{Z}) \to \mathrm{Aut}(M_{g'}) \cong \mathbb{C}/\Lambda ,$$

and one can obtain many examples of foliations with dense leaves.

Let $C = \mathbb{C}/\Lambda$ be an elliptic curve. Let X be the meromorphic vector field on the Riemann sphere associated to the differential equation

$$\frac{dz}{dT} = R(z) \qquad (T, z \in \mathbb{C})$$

where R is a rational function of degree greater that one with simple poles and zeroes. Let $\tilde{Y}$ be a constant nonzero vector field of $\mathbb{C}$ and let Y be the induced vector field on $\mathbb{C}/\Lambda$. Let $Z:\mathbb{C} \times C \to T(\mathbb{C} \times C)$ be the vector field defined by

$$Z(x,y) = (X(x), Y(y)) .$$

There exists a holomorphic line bundle tangent to $\hat{\mathbb{C}} \times C$ such that $Z(x,y)$ belongs to this line bundle, when x is not a pole of X. The induced foliation of the line bundle has leaves of mixed type. Let $p:\hat{\mathbb{C}} \times C \to \hat{\mathbb{C}}$ be projection onto the first factor and let $\{z_1,\ldots,z_{2n}\}$ be the set of zeroes and poles of X. Then, $\{z_1\} \times C$ $(1 \leq i \leq 2n)$ are leaves of the foliation. The foliation in $(\hat{\mathbb{C}} \times C) - \cup p^{-1}(z_i)$ is isomorphic to a foliation obtained by suspending a representation.

$$\Psi: \Pi_1(\hat{\mathbb{C}} - \{z_1,\ldots,z_{2n}\}) \to \mathrm{Aut}(C) \simeq \mathbb{C}/\Lambda$$

This is so because p restricted to a leaf in $\hat{\mathbb{C}} \times C - \cup p^{-1}(z_i)$ is a covering of the punctured sphere $\hat{\mathbb{C}} - \{z_1,\ldots,z_{2n}\}$. Since $n > 1$, every such leaf must be hyperbolic (I owe this example to A. Douady).

4) To a real quadratic field one can associate the Hilbert modular group of automorphisms of $\Delta \times \Delta$. This group, Γ, acts freely, properly and discontinuously. The quotient space $\Delta \times \Delta/\Gamma$ is a noncompact complete hyperbolic manifold. The action of Γ permutes the leaves of the foliation induced by projection onto the first factor and it also permutes the leaves of the foliation induced by projection onto the second factor. If L is a leaf of either foliation, then $\bigcup_{g \in \Gamma} g(L)$ is dense in $\Delta \times \Delta$. Hence, $\Delta \times \Delta/\Gamma$ has two transverse foliations F_1 and F_2, by Riemann surfaces where both F_1 and F_2 are minimal. One such example corresponds to $\mathbb{Q}(\sqrt{2})$. Let $\Gamma = PSL(2,\mathbb{Z}(\sqrt{2}))$ and for $g \in \Gamma$ let $\bar{g}$ denote the image of g under $\sigma:\Gamma \to \Gamma$, where σ is induced by the nontrivial Galois automorphism $a + b\sqrt{2} \longrightarrow a - b\sqrt{2}$. Then Γ acts on $\Delta \times \Delta$ as follows: $g \in \Gamma$, $g(x,y) = (g(x), \bar{g}(y))$. A subgroup of finite index $\tilde{\Gamma} \subset \Gamma$ acts without fixed points and in fact freely and properly.

If Γ is the Hilbert modular group of a real quadratic field, then $\Delta \times \Delta/\Gamma$ can be compactified by adding a certain number of cusps and the number of such cusps is equal to the class number of the field. Each cusp has a neighborhood which is homeomorphic to the cone over a 3-manifold which is a torus bundle over S^1. The ergodic properties of the horocyclic flows in the circle unit tangent bundles to these foliations are of very deep number-theoretical significance.

More generally, associated to a totally real algebraic number field of degree $n > 2$, there exists a complex complete hyperbolic manifold with n pairwise transverse foliations and these manifolds can be compactified by adding a certain number of cusps, which have neighborhoods which are cones over real $(2n-1)$-manifolds that are T^{2n-2} bundles over S^1.

§3. UNIFORMIZATION OF HYPERBOLIC FOLIATIONS. Let M be a complex manifold and let F be the holomorphic foliation which is obtained by integrating the complex line bundle $\eta \subset TM$. For $x \in M$ let η_x be the fibre at the point x. Let $\mathcal{D}_F(x) := \mathcal{D}(x) = \eta(x) \cap K(x)$, where $K(x)$ is the Kobayashi convex set at x.

(3.1) DEFINITION. We say that F is a hyperbolic foliation if $\mathcal{D}(x)$ is bounded for all $x \in M$. This implies that every leaf of F is hyperbolic.

If F is a hyperbolic foliation then $\mathcal{D}(x)$ is a round disk (balanced and absorbent). Then the Kobayashi Finsler structure, restricted to η, is a continuous hermitian metric on η. This hermitian metric will be called the *intrinsic* hermitian metric on η. This metric is, in general, only continuous.

(3.2) REMARK. Let M be a complex manifold and let F be a hyperbolic foliation, then one can always extend the hermitian intrinsic metric defined on the tangent line field of F to a continuous hermitian metric defined on all of M. However it is sometimes impossible to extend to a complete hermitian metric.

(3.3) DEFINITION. The hyperbolic foliation F is called a *complete* hyperbolic foliation if its intrinsic hermitian metric can be extended to a complete hermitian metric on the manifold on which F is defined (This condition is automatically satisfied if the manifold is a complete hyperbolic manifold).

For what follows we will need the following result of Van Dantzig and Van der Waerden (see Kobayashi [K] page 73).

(3.4) THEOREM. *Let* M *be a connected locally connected separable space with pseudo-distance* d_M. *Let* N *be a connected and locally compact complete metric space with distance* d_N. *The set* F *of distance decreasing mappings* $f: M \to N$ *is locally compact with respect to the compact-open topology. In fact, if* $C \subset N$ *is a compact subset of* N, *then the set* $F(x,C) = \{f \in F; f(x) \in C\}$ *is a compact subset of* F.

(3.5) COROLLARY. Let M be a complete hyperbolic complex manifold and let $r \geq 0$. Let $C_h^r(\Delta,M)$ denote the space of holomorphic mappings $f: \Delta \to M$ with the topology of uniform convergence, on compact sets, of all derivatives of order j, $0 \leq j \leq r$. Then $C_h^r(\Delta,M)$ is locally compact and the set $C_h^r(x,C,M,) = \{f \in C_h^r(\Delta,M); f(x) \in C\}$ is compact for each $x \in \Delta$ and compact subset $C \subset M$. In particular if M is compact, then $C_h^r(\Delta,M)$ is compact.

Let F be a complete hyperbolic foliation on the complex manifold M. Let d be the distance induced by a complete hermitian metric which extends the intrinsic metric of F. *Let* $F(F)$ be the family of holomorphic mappings $f:\Delta \to M$ such that $f(\Delta)$ is contained in some leaf of F. With this notation we have:

(3.6) PROPOSITION. $F(F)$ is closed in $C^r(\Delta,M)$ and it is locally compact for all $r \geqslant 0$. If M is compact then $F(F)$ is compact.

PROOF. Clearly, $F(F)$ is closed in $C^r(\Delta,M)$. For any $x,y \in \Delta$ and any $f \in F(F)$ we have

$$d(f(x),\ f(y)) \leqslant d_F(f(x),\ f(y)) \leqslant \rho(x,y)$$

where d denotes the extended distance and d_F is the distance along the leaves: if x and y belong to the same leaf then its distance along the leaves is the infimum of the lengths of piecewise differentiable curves, contained in the leaf with end points x,y. The proposition follows from Theorem (3.4).‖

Let M be a complex manifold with a complete hyperbolic foliation F. Let $\|\cdot\|: TM \to \mathbb{R}$ be the norm induced by a complete hermitian metric which extends the intrinsic metric.

(3.7) DEFINITION. The *modulus of uniformization* of F is defined as follows. Let $z \in M$ and let

$$\lambda_F(z) := \lambda(z) = \|\dot{u}_z(0)\|$$

where $u_z: \Delta \to L_z \subset M$ is a uniformization of L_z, the leaf through z, which satisfies $u_z(0) = z$. By the lemma of Schwarz, any uniformization $u: \Delta \to L_z$ which satisfies $u(0) = z$, is the form $u(z) = u_z(e^{i\theta}z)$, $\theta \in \mathbb{R}$. Therefore, $\lambda(z)$ does not depend on the chosen uniformization. Also, since $\dot{u}_z(0)$ is tangent to the foliation, it follows that $\lambda(z)$ does not depend on the extension of the intrinsic hermitian metric but only in the intrinsic metric itself. The function $\lambda: M \to \mathbb{R}^+$ is what we call the modulus of uniformization.

The following position is evident:

(3.8) PROPOSITION. For every $z \in M$, $0 < \lambda(z) \leqslant 1$.

(3.9) PROPOSITION. (Extremal Property). $\lambda: M \to \mathbb{R}^+$ satisfies the following extremal property:

$$(3.10)\quad \lambda(z) = \operatorname*{Max}_{\lambda \in F_z(F)} \{\|\dot{\phi}(0)\|\},$$

where $F_z(\mathcal{F}) \subset F(\mathcal{F})$ consists of all $\phi \in F(\mathcal{F})$ such that $\phi(0) = z$.

PROOF. The proof is immediate from Schwarz lemma. If $\phi \in F_z(\mathcal{F})$, let $\tilde{\phi}: \Delta \to \Delta$ be a lifting of ϕ with respect to a uniformization $u_z: \Delta \to L_z$, $u_z(0) = z$; such that $\tilde{\phi}(0) = 0$:

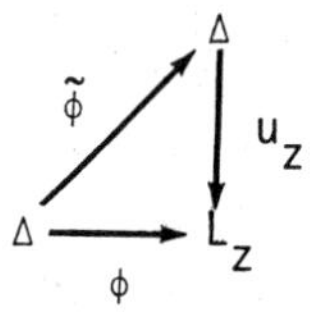

, $\tilde{\phi}(0) = 0$

By the lemma of Schwarz, $|\dot{\tilde{\phi}}(0)| \leqslant 1$ and ϕ is a uniformization if and only if $\tilde{\phi}(t) = e^{i\theta}T$ for some $\theta \in \mathbb{R}$. Hence, if $\phi \in F_z(\mathcal{F})$, $\|\dot{\phi}(0)\| \leq \lambda(z)$ and equality occurs if and only if ϕ is a uniformization. Since every leaf can be uniformized, the maximum is reached. //

(3.11) REMARK. If $\|\cdot\|_1$ is any hermitian norm on the tangent line field of $\mathcal{F}$, it induces on each leaf a metric which is conformal with respect to the intrinsic metric. The same proof of proposition 3.9 works as well to show that uniformizations are extremal of norms of derivatives of elements of $F(\mathcal{F})$ for *any* hermitian metric.

If $z_n \to z$ and $u_{z_n}: \Delta \to L_{z_n}$, $u_{z_n}(0) = z_n$ is the corresponding sequence of uniformizations, then there exists a subsequence which is uniformly convergent on compact subsets (by proposition 3.6). This fact plus the extremal property implies:

(3.12) COROLLARY. λ is lower semicontinuous.

(3.13) MAIN THEOREM. Let M be a complex n-manifold and let $\mathcal{F}$ be a complete hyperbolic foliation. Then $\lambda: M \to \mathbf{R}^+$ is a continuous function. Clearly, λ is a biholomorphic invariant of the pair $(M, \mathcal{F})$ in the sense that if $f: (M, \mathcal{F}) \to (N, \mathcal{F}')$ denotes a biholomorphism from M onto N which takes $\mathcal{F}$ onto $\mathcal{F}'$, then $\lambda(f(x)) = \lambda(x)$.

PROOF. By corollary 3.12, we only have to prove that λ is upper semicontinuous. Suppose the contrary. Then there exists a sequence $\{z_n\}_{n=1}^{\infty}$ such that $z_n \to z$ and σ, $0 < \sigma < 1$, such that

$$\lambda(z_n) < \lambda(z) - \sigma$$

for all $n \geqslant 1$. Let $\phi : \bar{\Delta} \to L_z$ be the mapping from the closed unit disk into L_z defined by

$$\phi(T) = u_z((1 - \frac{\sigma}{2})T), \quad |T| \leqslant 1$$

where $u_z : \Delta \to L_z$ is a uniformization such that $u_z(0) = z$. Then

$$(3.14) \quad \|\dot{\phi}(0)\| = (1 - \frac{\sigma}{2})\lambda(z) \geq \lambda(z_n) + \frac{\sigma}{2} \quad \text{for all} \quad n \geq 1 .$$

One may assume, by proposition 3.6, that one has a sequence of uniformization $u_{z_n} : \Delta \to L_z$, $u_{z_n}(0) = z_n$; such that $\{u_{z_n}\}$ is uniformly convergent on compact sets to a function $h : \Delta \to L_z$ which is *not* a uniformization of L_z. For any sequence $\{w_n\}$ such that $w_n \in L_{z_n}$ and $w_n \to w \in L_z$, there exists $\sigma' > 0$ such that $\lambda(w_n) < \lambda(w) - \sigma'$ for all sufficiently large n. Let $D^{2n-2} \subset \mathbf{R}^{2n-2}$ be the closed unit disk and let

$$g : \bar{\Delta} \times D^{2n-2} \to M$$

be a C^∞ submersion such that

i) $g|_{\bar{\Delta} \times \{0\}} = \phi$

ii) $g(\bar{\Delta} \times \{p\}) \subset L_{g(0,p)}$, for all $p \in D^{2n-2}$

iii) $g(\bar{\Delta} \times \{p\}) \cap g(\Delta \times \{q\}) = \phi$ if $p \neq q$.

Such submersion exists since $\bar{\Delta}$ is contractible and, therefore, the restriction of ϕ to any simple closed curve in $\bar{\Delta}$ represents the trivial element in the holonomy group of L_z. One can assume, without loss of generality that the sequence $\{z_n\}$ is of the form $\{g(0,p_n)\}$. For each $p \in D^{2n-2}$ let $g_p : \bar{\Delta} \to L_{g(0,p)}$ be defined by $g_p(T) = g(T, p)$, $(|T| < 1)$. Let $u_p : \bar{\Delta} \to L_{g(0,p)}$ be a uniformization such that $u_p(0) = g(0,p)$. Define $\phi_p : \bar{\Delta} \to \Delta$ as a lifting of g_p, with $\phi_p(0) = 0$:

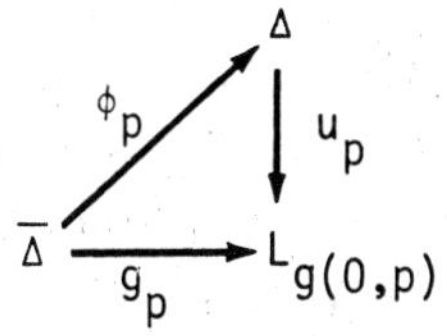

$\phi_p(0) = 0$

(3.15) LEMMA. The family $\{\phi_p\} \subset C^r(\overline{\Delta}, \Delta)$ is compact in the C^r topology of C^r mappings $f:\overline{\Delta} \to \Delta$, for all $r \geq 0$.

PROOF. Since $\overline{\Delta} \times D^{2n-2}$ is compact, g is smooth, and u_p is conformal, it follows that there exists $K > 0$ such that ϕ_p is K-quasiconformal for every p. Therefore, the family $\{\phi_p\}$ is locally compact in the C^r topology (See O. Letho, K.I. Virtanen [L-V].) Since the family fixes the origin it is in fact compact. //.

Now let us consider the sequence $\{\phi_{p_n}\}$. One may assume that this sequence converges uniformly to a function $h: \overline{\Delta} \to \Delta$, such that $h(0) = 0$. But this function must be holomorphic, since $g_0 = \phi$ and therefore, the complex dilations $\mu_{p_m} : \overline{\Delta} \to \mathbb{R}$, of ϕ_{p_m} tend uniformly to zero as $n \to \infty$. Since the limit of these complex dilations is the complex dilation of h, it follows that h is holomorphic. On the other hand, by 3.14, and since $\dot{g}_{p_m}(0) \to \dot{\phi}(0)$ as $n \to \infty$ it follows that $|\dot{\phi}_{p_m}(0)| \geq 1 + \frac{\sigma}{3}$ for n sufficiently large. Hence, $|\dot{h}(0)| \geq 1 + \frac{\sigma}{3}$. This contradicts Schwarz lemma. Hence, our main theorem (3.13) is true. //

For all the rest of §3, F will be a complete hyperbolic foliation on M.

(3.16) COROLLARY. Let $U(F)$ denote the set of all holomorphic mappings $u:\Delta \to M$ such that u is a uniformization of some leaf of F. Then $U(F)$ is locally compact in $C^r(\Delta, M)$ for all $r \geq 0$. If M is compact, then $U(F)$ is compact. Therefore, limits of uniformizations of leaves are uniformizations of leaves. We will call $U(F)$ the *uniformization family* of F.

If $u \in U(F)$ and $u:\Delta \to L$, then any other uniformization of L is obtained by precomposing u with an element g, of $PSL(2, \mathbb{R}) = Aut(\Delta)$.

One has a natural action of $PSL(2, \mathbb{R})$ on $U(F)$ as follows:

$$g(u) = u \circ g \ ; \ g \in PSL(2, \mathbb{R}), \ u \in U(F).$$

This action is locally free. We have a natural projection $\Pi:U(F) \to M$ defined by $\Pi:U(F) \to M$ defined by $\Pi(u) = u(0)$. If $v_z \neq 0$ is vector which is tangent to L_z at z, then there exists a unique uniformization $u_z:\Delta \to L_z$ such that $\dot{u}_z(0) = \mu v_z$, $\mu > 0$. Hence, $U(F)$ can be canonically identified with the unit circle bundle tangent to F. With this identification, $PSL(2, \mathbb{R})$ acts transitively on the unit tangent bundle of each leaf.

(3.17) COROLLARY. The natural action of $PSL(2, \mathbb{R})$ on $U(F)$ is continuous and locally free. Furthermore, $\Pi: U(F) \to M$ is a continuous locally trivial fibration with fibre S^1. The isotropy subgroups are the discrete subgroups of $PSL(2, \mathbf{R})$ which are the fundamental group of the unit tangent bundles of the leaves of F. These isotropy subgroups vary upper semicontinuously as closed subsets of $PSL(2, \mathbf{R})$ (in the sense that $\{(g, u) \in PSL(2, \mathbf{R}) \times U(F); g(u) = u\}$ is a closed subset of $PSL(2, \mathbf{R}) \times U(F)$).

(3.18) If M is a compact hyperbolic manifold then none of the leaves has cusps, since any holomorphic mapping from a punctured disk into M extends to the whole disk (See Corallary 3.2, pag. 83 [K]). Therefore the isotropy groups in Corollary 3.17 can not contain parabolic elements.

If $L \in F$ then its *limit set*, $L(L)$, is the set of all points in M such that there exists a sequence of points in the leaf which converges to the point and such that the sequence is not contained in any compact subset of the leaf.

(3.19) COROLLARY. Let $w \in L(L_z)$ then there exists a sequence $\{g_i\}_1^\infty$, $g_i \in PSL(2, \mathbf{R})$ such that $\{u_z \circ g_i\}$ converges to a uniformization, u of L_w such that $u(0) = w$. In particular, if L is a dense leaf then L determines all the information, M and F, in a equicontinuous way.

One way to obtain complete hyperbolic foliations is to take a complete hyperbolic open region of $\mathbb{C}^n$ and restrict a nonsingular foliation of $\mathbb{C}^n$ by curves, If $\Omega \subset \mathbb{C}^n$ is a bounded convex region whose boundary is not smooth then the intrinsic metric of a foliation might be nonsmooth even along the leaves. The modulus of uniformization gives a natural and intrinsic hermitian metric on the bundle which defines the foliation:

(3.20) DEFINITION. Let η denote the tangent bundle of F. Let $\| \ \|$ denote its intrinsic norm, and let $\lambda: M \to \mathbf{R}$ be its modulus of uniformization. The *foliated intrinsic hemitian metric*, $\| \ \|_*: \eta \to \mathbf{R}$ is defined by $\|V_x\|_* = (\lambda(x))^{-1} \|V_x\|$, where $x \in M$, $V_x \in \eta_x$. This metric can always be extended to a complete hermitian metric in all of TM. Thus, this metric has the property that any uniformization $u: \Delta \to L_z$ sends Δ isometrically, when Δ has the Poincaré metric.

With this notation we have the following Corollary of 3.13:

(3.21) COROLLARY. The foliated intrinsic hermitian metric is continuous and real analytic along the leaves. It induces a metric of constant negative curvature equal to minus one on each leaf. Given $\delta > 0$, there exists a smooth and complete hermitian metric on M such that $-1 - \delta < k(x) < -1 + \delta$, where $k(x)$ denotes the gaussian curvature of L_x, at the point x, in this metric.

This metric can be taken to be an arbitrarly small smooth approximation of a hermitian metric which is complete and extends the foliated intrinsic metric. If the foliated intrinsic metric is smooth, then there exists a smooth hermitian metric in M such that every leaf has constant negative curvature equal to minus one.

Let N be the unit circle bundle of η with respect to the foliated intrinsic metric. By 3.16 and 3.17 there exists a locally free continuous action of $PSL(2, \mathbb{R})$ on N. The action is real analytic, when restricted to the unit circle bundle of a leaf. Let X, Y and Z be the left-invariant vector fields of $SL(2, \mathbb{R})$ corresponding to the one-parameter subgroups

$$(3.22) \quad t \to \begin{bmatrix} e^t & 0 \\ 0 & e^{-t} \end{bmatrix}, \quad t \to \begin{bmatrix} 1 & t \\ 0 & 1 \end{bmatrix}, \quad t \to \begin{bmatrix} 1 & 0 \\ t & 1 \end{bmatrix},$$

respectively. Via the double covering $SL(2, \mathbb{R}) \to PSL(2, \mathbb{R})$, we see that these one-parameter subgroups induce three continuous flows: $g_t : N \to N$, $h_t^+ : N \to N$, $h_t^- : N \to N$ These are the *foliated geodesic flow, foliated unstable hocyclic flow* and *foliated stable horocyclic flow,* respectively. These are the flows we used in proposition 2.15.

§4. HOLOMORPHIC VECTOR FIELDS NEAR AN ISOLATED SINGULARITY. Let us consider the differential equation

$$(4.1) \quad \frac{dz}{dT} = F(z), \quad (T \in \mathbb{C}, \quad z \in \mathbb{C}^n)$$

defined in the open set $U \subset \mathbb{C}^n$ and given by the holomorphic vector field $F: U \to \mathbb{C}^n$. The local solutions will be holomorphic functions $\phi : V \subset \mathbb{C} \to U$. If $S = F^{-1}(\{0\})$ is the singular set of F, then the local solutions of 4.1 fit together and give raise to a holomorphic foliation, by Riemann surfaces, of $U - S$. Since we are interested in the local behavior of solutions near an isolated singularity, one may assume without loss of generality that $S = \{0\}$ and U is the open unit ball of $\mathbb{C}^n$. In order to apply our previous results it will be important to provide the unit ball D^n with the Poincaré-Bergman metric:

$$(4.2) \quad dS^2 = \frac{4\|dz\|^2}{(1-\|z\|^2)^2} \; ; \; \|z\|^2 = \sum_{i=1}^{n} z_i \bar{z}_i \; , \; \|dz\|^2 = \sum_{i=1}^{n} dz_i \, \overline{dz}_i \, , \; z \in D^n .$$

This choice is not a minor point since we need the domain of definition of F to be a complete hyperbolic manifold and the unit ball, being a bounded symmetric domain, seems like a good choice for the local analysis of vector fields.

One important property of nonsingular solutions of 4.1 is that they have negative gaussian curvature with respect to the metric induced by the euclidean metric of $\mathbb{C}^n$. If F is not the identity map (i.e., if F is not the "radial" vector field $F(z)=z$) and if 0 is an isolated singularity of F, then the curvature accumulates towards $-\infty$ as we get closer to the origin. Let $\langle , \rangle$ denote the standard hemitian metric of $\mathbb{C}^n$ and $\| \ \|$ the respective norm. Let $\rho : U \to \mathbb{R}$ be a smooth positive function and let us consider the metric

$$(4.3) \quad dS^2 = \rho^2 \| dz \|^2 = \rho^2 (|dz_1|^2 + \ldots + |dz_n|^2) .$$

Consider the function $K_{F,\rho}(w) := K(w)$, $w \in U - S$, where $K(w)$ is the gaussian curvature of the solution of 4.1, passing through w, with respect to the conformal metric 4.3. We would like to compute K using the fact that we have a differential equation. Let $w \in U - S$ and let $\phi : V \to \mathbb{C}^n$ be a local solution of 4.1 passing through w, where $V \subset \mathbb{C}$ is an open set containing the origen:

$$(4.4) \quad \begin{cases} \dot{\phi}(T) = F(\phi(T)), \ T \in V \\ \phi(0) = w \end{cases}$$

The metric induced by ϕ on V from the metric 4.3 is conformal with respect to the euclidean metric, $|dT|^2$, on the plane:

$$(4.5) \quad ds^2 = \mu^2 |dT|^2 ,$$

where $\mu(T) = [\rho(\phi(T))] \|F(\phi(T))\|$, $T \in V$.

Therefore,

$$(4.6) \quad K(w) = -[\rho(\phi(0))]^{-2} \|F(\phi(0))\|^{-2} [\tfrac{1}{4} \tfrac{\partial}{\partial T} \tfrac{\partial}{\partial \bar{T}} (\log((\rho \circ \phi) \|F \circ \phi\|)]_{T=0} =$$

$$= -[\rho(w)]^{-2} \|F(w)\|^{-2} h(w) .$$

From the chain rule and the fact that ϕ is a solution of 4.1 it follows that the curvature depends only on the right-hand of the differential equation. It is a nice fact that to compute K we do not need to integrate the differential equation.

Now let $U = D^n$ with the Poincaré-Bergman metric 4.2. Let $\phi : V \to D^n$ be a non-constant solution of 4.1. Then, we have:

$$(4.7) \quad \begin{cases} \text{i)} \ \frac{\partial}{\partial T} F(\phi(T)) = DF_{\phi(T)}(F(\phi(T))) = \ddot{\phi}(T) \\ \text{ii)} \ \frac{\partial}{\partial T} \overline{\phi(T)} = 0 , \ \frac{\partial}{\partial \bar{T}} \phi(T) = 0 \\ \text{iii)} \ \frac{\partial}{\partial T} \overline{F(\phi(T)} = 0 , \ \frac{\partial}{\partial \bar{T}} F(\phi(T)) = 0 \end{cases}$$

Let $K_h: D^n - S \to \mathbb{R}$ denote the curvature function with respect to the metric 4.2. The we have

(4.8) PROPOSITION:

$$K_h(z) = \|z\|^2 - 1 - |<z, \frac{F(z)}{\|F(z)\|}>|^2 -$$

$$- \frac{(1-\|z\|^2)^2 [\|F(z)\|^2 \|DF_z(F(z))\|^2 - |<F(z), DF_z(F_z)>|^2]}{2\|F(z)\|^6}$$

Notice that by Schwartz inequality we have $K_h(z) < 0$. If λ is a non-vanishing holomorphic function one checks directly that the formula (4.8) gives the same value for the vector field λF. When F is the "radial" vector field $F(z) = z$, then $K_h(z) \equiv -1$.

If we had taken the euclidean metric $\|dz\|^2$, then the euclidean curvature function $K_e: D^n - S \to \mathbb{R}$ is given by the following formula:

$$(4.9) \quad K_e(z) = - \frac{\|F(z)\|^2 \|DF_z(F(z))\|^2 - <F(z), DF_z(F(z))>^2}{\|F(z)\|^6}$$

Formulas 4.8 and 4.9 imply that, in general, there exists a sequence $z_n \to 0$ such that $K_h(z_n) \to -\infty$ and the same is true for K_e.

An important problem for holomorphic vector fields near an isolated singularity is to characterize the nature of solutions that "approach" the origin. A *local separatrix* is the germ of an analytic curve, V, at the origin, such that $V - \{0\}$ is tangent to the vector field. We are assuming that the origin is an isolated singularity and V is nonsingular outside of the origin. The problem of the existence of a separatrix through an isolated singularity was proposed by Briot and Bouquet in 1856. It has been shown by C. Camacho and P. Sad [C - S] that in complex dimension 2 there exists always a separatrix through an isolated singularity. In a sense separatrices can be thought as orbits of 4.1 that approach the origin in the nicest way. The separatrix problem remains open for $n > 2$.

Let $F: D^n \to \mathbb{C}^n$ be a holomorphic vector field with an isolated singularity at the origin. We provide D^n with the Poincaré-Bergman metric and we provide each leaf of the foliation, $\mathcal{F}(F)$, determined by F in $D^n - \{0\}$ with the induced metric. For all that follows we will assume these hypothesis.

(4.10) DEFINITION. A leaf $L \in \mathcal{F}(F)$ is called a *quasi-separatrix* if L is incomplete with respect to the metric induced on L by the Poincaré-Bergman metric. Clearly, every separatrix is a quasi-separatrix.

The following proposition is immediate from the previous definition.

(4.11) PROPOSITION. L is a quasi-separatrix if and only if there exists a geodesic of the leaf, $\alpha:[0,d) \to L$, parametrized by arc length, $d < \infty$, such that.

$$\lim_{t \to d} \alpha(t) = 0 .$$

Motivated by the separatrix problem we propose the following:

(4.12) QUESTION. Do pseudo-separatrices always exist at an isolated singularity?

Let $F: D^n \to \mathbb{C}^n$ be holomorphic and $F^{-1}(\{0\}) = \{0\}$. Then, there exists $0 < \delta < 1$ such that for all $z_0 \in D^n_\delta$, where D^n_δ is the open ball of radius δ, there exists a local solution $\phi: \Delta \to D^n$, defined on the unit disk, of the initial value problem

$$\frac{dz}{dT} = F(z) \tag{4.13}$$
$$\phi(0) = z_0 .$$

This solution parametrizes an open neighborhood of $z_0 \in L_{z_0}$, if $z_0 \neq 0$.

Since uniformizations are extremals of functions from the disk to the leaves, it seems interesting to compare the solution of 4.13 with a uniformization $U_{z_0}: \Delta \to D^n$; $U_{z_0}(0) = z_0$. Let us lift ϕ by the uniformization U_{z_0}:

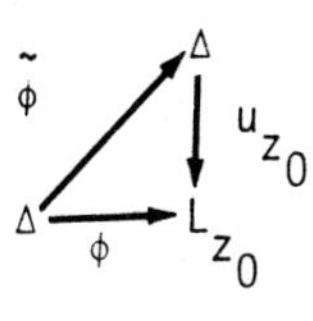

, $\tilde{\phi}(0) = 0$.

By Schwarz lemma, one has $|\dot{\tilde{\phi}}(0)| \leq 1$.

Define $\lambda_F: D^n_\delta - \{0\} \to \mathbf{R}$ as follows:

$$\lambda_F(z_0) = \frac{\| \dot{U}_{z_0}(0) \|}{| F(z_0) |}$$

Since $|\dot{\tilde{\phi}}(0)| \leqslant 1$ one has:

(4.14) PROPOSITION. $\lambda_F(z_0) \geqslant 1$, for all $z_0 \in D^n_\delta - \{z_0\}$.

(4.15) COROLLARY. Let $z_n \to w$ be a sequence of points in $D^n_\delta - \{0\}$ which converges to $w \in D^n_\delta - \{0\}$. Let $\{U_{z_n}\}$ be a sequence of uniformizations of leaves of $\mathcal{F}(F)$ such that $U_{z_n}(0) = z_n$. Then $\{U_{z_n}\}$ contains a subsequence which converges uniformly on compact subsets to a nonconstant holomorphic function $u: \Delta \to D^n$. Furthermore, $u(\Delta) \subset L_w \cup \{0\}$. Hence if $u(\Delta)$ contains the origin then it is a separatrix.

PROOF OF COROLLARY. Since D^n is a complete hyperbolic manifold and $\| U_{z_n}(0)\| < \delta < 1$, it follows that $\{U_{z_n}\}$ contains a convergent subsequence. Therefore, the limit of any convergent subsequence is a nonconstant holomorphic function. Clearly, any such limit, U, satisfies $U(\Delta) \subset L_w \cup \{0\}$ //.

If no limit contains the origin in corollary 4.15, one can apply exactly the same arguments as in the proof of Theorem 3.13 to prove the following:

(4.16) PROPOSITION. If $F: D^n \to \mathbb{C}^n$ is a holomorphic vector field with $F^{-1}(\{0\}) = \{0\}$ and if this vector field does not have a separatrix, then $\lambda_F: D^n - \{0\} \to \mathbb{R}$ is continuous and can be extended as a continuous function to all of D^n defining $\lambda_F(0) = 0$.

(4.17) EXAMPLE 1). If $f: D^2 \to \mathbb{C}$ is a holomorphic function with $f(0)=0$ and if 0 is the only singular point of f, then one has a foliation, defined by the level surfaces of f, on D^2. This function f is a first integral for the holomoprphic vector field in D^2:

$$\frac{dz_1}{dT} = -\frac{\partial f}{\partial z_2}, \quad \frac{dz_2}{dT} = \frac{\partial f}{\partial z_1}$$

and $f^{-1}(0)$ is a separatrix of this vector field. One can give examples of functions f such that $\lambda_F: D^2 - \{0\} \to \mathbf{R}$ is not continuous, where F is a vector field that has f as a first integral.

2) If A is a complex $(n \times n)$-matrix such that its eigenvalues $\lambda_1, \ldots, \lambda_n$ have the property that 0 is not in their convex envelope then one says that A belongs to the Poincaré Domain. Let A belong to the Poincaré Domain and let us consider the complex linear differential equation in $\mathbb{C}^n$:

$$(4.18) \quad \frac{dz}{dT} = Az.$$

Equation 4.18 defines a complex flow $\phi_T: \mathbb{C}^n \to \mathbb{C}^n$, $T \in \mathbb{C}$. The hypothesis that A belongs to the Poincaré domain implies that there exists $a \in \mathbb{C}^*$ such that the real flow $f_t: \mathbb{C}^n \to \mathbb{C}^n$, $t \in \mathbf{R}$, defined by $f_t(z) = \phi_{ta}(z)$ has the origin as a hyperbolic attractor. Then, every leaf of the nonsingular foliation induced by 4.18 in the unit disk $D^n\{X\}$, will interesect each sphere S_δ^{2n-1}, $0 < \delta < 1$, transversely. Hence, every is a pseudo-separatrix.

BIBLIOGRAPHY

[A-A] Arnold V., Avez A. "Problemès ergodiques de la mecanique classique". Gauthier-Villars (Paris) 1967.

[A-B] Ahlfors L., Bers L. "Riemann's mapping theorem for variable metrics". Ann. of Math, 72 (1960) 385-404.

[C-S] Camacho C., Sad P. "Invariant varietes through singularities of holomorphic vector fields". Ann. of Math., Vol. 115 No. 3 (1982) 579-596.

[J] Jørgensen T. "Compact 3 manifolds of constant negative curvature". Ann. of Math. (106), 1977, 61-72.

[K] Letho, O. Virtanen, K.I. "Quasi-conformal mappings in the plane". Springer-Verlag, Berlin, 1973.

[T] Thurston, W. "The geometry and topology of the three-manifolds", Princeton University, Math. Dept. 1979.

DEPARTMENT OF MATHEMATICS
CENTRO DE INVESTIGACION Y DE
ESTUDIOS AVANZADOS DEL IPN
APARTADO POSTAL 14-740,
MEXICO 14, D.F. 07000 - MEXICO

Copying and reprinting. Individual readers of this publication, and nonprofit libraries acting for them, are permitted to make fair use of the material, such as to copy an article for use in teaching or research. Permission is granted to quote brief passages from this publication in reviews provided the customary acknowledgement of the source is given.

Republication, systematic copying, or multiple reproduction of any material in this publication (including abstracts) is permitted only under license from the American Mathematical Society. Requests for such permission should be addressed to the Executive Director, American Mathematical Society, P.O. Box 6248, Providence, Rhode Island 02940.

The appearance of the code on the first page of an article in this volume indicates the copyright owner's consent for copying beyond that permitted by Sections 107 or 108 of the U. S. Copyright Law, provided that the fee of $1.00 plus $.25 per page for each copy be paid directly to Copyright Clearance Center, Inc., 21 Congress Street, Salem, Massachusetts 01970. This consent does not extend to other kinds of copying, such as copying for general distribution, for advertising or promotion purposes, for creating new collective works or for resale.

ABCDEFGHIJ — 8987